Natural Language Processing and Computational Linguistics 1

Series Editor
Patrick Paroubek

Natural Language Processing and Computational Linguistics 1

Speech, Morphology and Syntax

Mohamed Zakaria Kurdi

WILEY

First published 2016 in Great Britain and the United States by ISTE Ltd and John Wiley & Sons, Inc.

ISTE Ltd
27-37 St George's Road
London SW19 4EU
UK

www.iste.co.uk

John Wiley & Sons, Inc.
111 River Street
Hoboken, NJ 07030
USA

www.wiley.com

Library of Congress Control Number: 2016945024

British Library Cataloguing-in-Publication Data
A CIP record for this book is available from the British Library
ISBN 978-1-84821-848-2

Contents

Introduction

Language is one of the central tools in our social and professional life. Among other things, it acts as a medium for transmitting ideas, information, opinions and feelings, as well as for persuading, asking for information, giving orders, etc. Computer Science began to gain an interest in language as soon as the field itself emerged, notably within the field of Artificial Intelligence (AI). The Turing test, one of the first tests developed to judge whether a machine is intelligent or not, stipulates that to be considered intelligent, a machine must possess conversational abilities that are comparable to those of a human being [TUR 50]. This implies that an intelligent machine must possess comprehension and production abilities, in the broadest sense of these terms. Historically, natural language processing (NLP) got itself focused on the potential for applying such technology to the real world in a very short span of time, particularly with machine translation (MT) during the Cold War. This began with the first machine translation system which was developed as the brainchild of a joint project between the University of Georgetown and IBM in the United States [DOS 55, HUT 04]. This work was not crowned with the success that was expected, as the researchers soon realized that a deep understanding of the linguistic system is a prerequisite for any comprehensive application of this kind. This discovery, presented in the famous report by automatic language processing advisory committee (ALPAC), had a considerable impact upon machine translation work and on the field of NLP in general. Today, even though NLP is largely industrialized, the interest in basic language processing has not waned. In fact, whatever the application of modern NLP, the use of a basic language processing unit such as a morphological, syntactic, recognition or speech synthesis analyzer is almost always indispensable (see [JON 11] for a more complete review of the history of NLP).

I.1. The definition of NLP

Firstly, what is NLP? It is a discipline which is found at the intersection of several other branches of science such as Computer Science, Artificial Intelligence and Cognitive Psychology. In English, there are several terms for certain fields which are very close to one another. Even though the boundaries between these designated fields are not always very clear, we are going to try to give a definition without claiming that the definition is unanimously accepted in the community. For example, the terms *formal linguistics* or *computational linguistics* relate more to models or linguistic formalities developed for IT implementation. The terms *Human Language Technology* or *Natural Language Processing*, on the other hand, refer to a publishing software tool equipped with features related to language processing. Furthermore, *speech processing* designates a range of techniques from signal processing to the recognition or production of linguistic units such as phonemes, syllables or words. Except for the dimension dealing with the signal processing, there is no major difference between speech processing and NLP. Many techniques that have initially been applied to speech processing have found their way into applications in NLP, an example being the Hidden Markov Models (HMM). This encouraged us to follow the unifying path already taken by other colleagues, such as [JUR 00], in this book. This path involves grouping NLP and speech processing into the same discipline. Finally, it is probably worth to mention the term *corpus linguistics* which refers to the methods of collection, annotation and use of corpora, both in linguistic research and NLP. Since corpora have a very important role in the process of constructing an NLP system, notably those which adopt a machine learning approach, we saw fit to consider corpus linguistics as a branch of NLP.

In the following sections, we will present and discuss the relationships between NLP and related disciplines such as linguistics, AI and cognitive science.

I.1.1. *NLP and linguistics*

Today, with the democratization of NLP tools, such tools make up the toolkit of many linguists conducting empirical work across a corpus. Therefore, Part-Of-Speech (POS) taggers, morphological analyzers and syntactic parsers of different types are often used in quantitative studies.

They may also be used to provide the necessary data for a psycholinguistics experiment. Furthermore, NLP offers linguists and cognitive scientists a new perspective by adding a new dimension to research carried out within these fields. This new dimension is testability. Indeed, many theoretical models have been tested empirically with the help of NLP applications.

I.1.2. *NLP and AI*

AI is the study, design and creation of intelligent agents. An intelligent agent is a natural or artificial system with perceptual abilities that allows it to act in a given environment to satisfy its desires or successfully achieve planned objectives (see [MAR 14a] and [RUS 10] for a general introduction). Work in AI is generally classified into several sub-disciplines or branches, such as knowledge representation, planning, perception and learning. All these branches are directly related to NLP. This gives the relationship between AI and NLP a very important dimension. Many consider NLP to be a branch of AI while some prefer to consider NLP a more independent discipline.

In the field of AI, planning involves finding the steps to follow to achieve a given goal. This is achieved based on a description of the initial states and possible actions. In the case of an NLP system, planning is necessary to perform complex tasks involving several sources of knowledge that must cooperate to achieve the final goal.

Knowledge representation is important for an NLP system at two levels. On the one hand, it can provide a framework to represent the linguistic knowledge necessary for the smooth functioning of the whole NLP system, even if the size and the quantity of the declarative pieces of information in the system vary considerably according to the approach chosen. On the other hand, some NLP systems require extralinguistic information to make decisions, especially in ambiguous cases. Therefore, certain NLP systems are paired with ontologies or with knowledge bases in the form of a semantic network, a frame or conceptual graphs.

In theory, perception and language seem far from one another, but in reality, this is not the case, especially when we are talking about spoken language where the linguistic message is conveyed by sound waves produced by the vocal folds. Making the connection between perception and voice recognition (the equivalent of perception with a comprehension element) is crucial, not only for comprehension, but also to improve the quality of speech recognition. Furthermore, some current research projects are looking at the connection between the perception of spoken language and the perception of visual information.

Machine learning involves building a representation after having examined data which may or may not have previously been analyzed. Since the 2000s, machine learning has gained particular attention within the field of AI, thanks to the opportunities it offers, allowing intelligent systems to be built with minimal effort compared to rule-based symbolic systems which require more work to be done by human experts. In the field of NLP, the extent to which basic machine learning is used depends highly on the targeted linguistic level. The extent to which machine learning is used varies between almost total domination within speech recognition systems and limited usage within high level processing such as in discourse analysis and pragmatics, where the symbolic paradigm is still dominant.

I.1.3. *NLP and cognitive science*

As with linguistics, the relationship between cognitive science and NLP goes in two directions. On the one hand, cognitive models can act to support a source of inspiration for an NLP system. On the other hand, constructing an NLP system according to a cognitive model can be a way of testing this model. The practical benefit of an approach which mimics the cognitive process remains an open question because in many fields, constructing a system which is inspired by biological models does not prove to be productive. It should also be noted that certain tasks carried out by NLP systems have no parallel in humans, such as searching for information across search engines or searching through large volumes of text data to extract useful information. NLP can be seen as an extension of human cognitive capabilities as part of a decision support system, for example. Other NLP systems are very close to human tasks, such as comprehension and production.

I.1.4. *NLP and data science*

With the availability of more and more digital data, a new discipline has recently emerged: data science. It involves extracting, quantifying and visualizing knowledge, primarily from textual and spoken data. Since these data are found in natural language in many cases, the role of NLP in the extraction and treatment process is obvious. Currently, given the countless industrial uses for this kind of knowledge, especially within the fields of marketing and decision-making, data science has become extremely important, even reminiscent of the beginning of the Internet in the 1990s. This shows that NLP is as useful when applied as it is when considered as a research field.

I.2. The structure of this book

The aim of this book is to give a panoramic overview of both early and modern research in the field of NLP. It aims to give a unified vision of fields which are often considered as being separate, for example speech processing, computational linguistics, NLP and knowledge engineering. It aims to be profoundly interdisciplinary and tries to consider the various linguistic and cognitive models as well as the algorithms and computational applications on an equal footing. The main postulate adopted in this book is that the best results can only be the outcome of a solid theoretical backbone and a well thought-out empirical approach. Of course, we are not claiming that this book covers the entirety of the works that have been done, but we have tried to strike a balance between North American, European and international work. Our approach is thus based on a duel perspective, aiming to be accessible and informative on the one hand but on the other, presenting the state-of-the-art of a mature field which is in a constant state of evolution.

As a result, this work uses an approach that consists of making linguistic and computer science concepts accessible by using carefully chosen examples. Furthermore, even though this book seeks to give the maximum amount of detail possible about the approaches presented, it nevertheless remains neutral about implementation details to leave each individual some freedom regarding the choice of a programming language. This must be chosen according to personal preference as well as the specific objective needs of individual projects.

Besides the introduction, this book is made up of four chapters. The first chapter looks at the linguistic resources used in NLP. It presents the different types of corpora that exist, their collection, as well as their methods of annotation. The second chapter discusses speech and speech processing. Firstly, we will present the fundamental concepts in phonetics and phonology and then we will move to the two most important applications in the field of speech processing: recognition and synthesis. The third chapter looks at the word level and it focuses particularly on morphological analysis. Finally, the fourth chapter covers the field of syntax. The fundamental concepts and the most important syntactic theories are presented, as well as the different approaches to syntactic analysis.

Linguistic Resources for NLP

Today, the use of good linguistic resources for the development of NLP systems seems indispensable. These resources are essential for creating grammars, in the framework of symbolic approaches or to carry out the training of modules based on machine learning. However, collecting, transcribing, annotating and analyzing these resources is far from being trivial. This is why it seems sensible for us to approach these questions in an introduction to NLP. To find out more about the matter of linguistic data and corpus linguistics, a number of works and articles can be consulted, including [HAB 97, MEY 04, WIL 06a, WIL 06b] and [MEG 03].

1.1. The concept of a corpus

At this point, a definition of the term *corpus* is necessary, given that it is central for the subject of this section. It is important to note that research works related to both written and spoken language data is not limited to corpus linguistics. It is actually possible to use individual texts for various forms of literary, linguistic and stylistic analyses. In Latin, the word corpus means *body*, but when used as a source of data in linguistics, it can be interpreted as *a collection of texts*. To be more specific, we will quote scholarly definitions of the term *corpus* from the point of view of modern linguistics:

– A collection of linguistic data, either written texts or a transcription of recorded speech, which can be used as a starting point of linguistic description or as a means of verifying hypotheses about a language [CRY 91].

– A collection of naturally occurring language text, chosen to characterize a state or variety of a language [SIN 91].

– The corpus itself cannot be considered as a constituent of the language: it reflects the character of the artificial situation in which it has been produced and recorded [DUB 94].

From these definitions, it is clear that a corpus is a collection of data selected with a descriptive or applicative aim as its purpose. However, what exactly are these collections? What are their fundamental properties? It is generally thought that a corpus must possess a common set of fundamental properties, including representativeness, a finite size and existing in electronic format.

The problem with the representativeness of a corpus has been highlighted by Chomsky. According to him, certain entirely valid linguistic phenomena exist which might never be observed due to their rarity. Given the infinite nature of language due to the possibility of generating an infinite number of different sentences from a finite number of rules and the constant addition of neologisms in living languages, it is clear that whatever be the size of a corpus, it would be impossible to include all linguistically valid phenomena. In practice, researchers construct corpora whose size is geared to the individual needs of the research project. Thus, the phenomena that Chomsky is talking about are certainly linguistically valid from a theoretical point of view but are almost never used in everyday life. A sentence that is ten thousand words long and formed in accordance with the rules of the English language is of no interest to a researcher who is trying to construct a machine translation system from English to Arabic, for example. Furthermore, we often talk about applications which are task orientated, where we are looking to cover the linguistic forms used in an applied context, which is restricted to hotel reservations or asking for tourist information, for example. In this sort of application, even though it is impossible to be exhaustive, it is possible (even though it takes a lot of work) to reach a satisfactory level.

Often, the size of a corpus is limited to the given number of words (a million words, for example). The size of a corpus is generally predetermined in advance during the design phase. Sometimes, teams, such as Professor John Sinclair's team at the University of Birmingham in England, update their corpus continuously (in this case, the term *text collection* is preferred). This continuous updating is necessary to guarantee the representativeness of a corpus across time: the opening up and the infinity of the corpus constitute a means to guarantee diachronic representativeness. Infinite corpora are particularly useful for lexicographers who are looking to include neologisms in new editions of their dictionaries.

Today, the word corpus is almost automatically associated with the word digital. Historically, the term referred mainly to printed texts or even manuscripts. The advantages of digitalization are undeniable. On the one hand, research has become much easier and results are obtained more quickly and, on the other hand, annotation can be done much more flexibly. Moreover, sometimes long-distance teamwork has become much easier. Furthermore, in view of the extreme popularity of digital technology, having data in an electronic format allows such data to be exchanged and allows paper usage to be reduced (which is a good thing given the impact of paper usage on the environment). However, this gave birth to some long-term issues related to electronic corpora such as portability. With the development of operating systems and text analysis software, it sometimes becomes difficult to access documents that were coded with old versions of software with a format that is obsolete. To get around this problem, researchers try to perpetuate their data using independent versions of platforms and of text processing software. XML markup language is one of the main languages used for the annotation of data. More specialized standards such as the EAGLES Corpus Encoding Standard and XCES are also available and are under continuous development to allow researchers to understand linguistic phenomena in a precise and reliable way.

In the field of NLP, the use of corpora is uncontested. Of course, there is a debate surrounding the place of corpora within the approach to build NLP systems, but to our knowledge, everyone is in agreement that linguistic data play a very important role in this process. Corpora are also very useful within linguistics itself, especially for those who wish to carry out a study on a specific linguistic phenomenon such as collocations, fixed expressions, as well as lexical ambiguities. Furthermore, corpora are used more and more in disciplines such as cognitive science or foreign language teaching [NES 05, GRI 06, ATW 08].

1.2. Corpus taxonomy

To establish a corpus taxonomy, many criteria can be used, such as the distinction between spoken corpora, written corpora, modern corpora, corpora of an ancient form of a language or a dialect, as well as the number of languages in a given corpus.

1.2.1. *Written versus spoken*

This kind of corpus is made up of a collection of written texts. Often, corpora such as these contain newspaper articles, webpages, blogs, literary or religious texts, etc. Another source of data from the Internet includes written dialogues between two people communicating on the Internet (such as in a chat) or between a person and a computer program designed specifically for this kind of activity. Often, newspaper archives such as *The Guardian* (for English), *Le Monde* (for French) and *Al-Hayat* (for Arabic) are also a very popular source for written texts. They are especially useful within the fields of information research and lexicography. More sophisticated corpora also exist, such as the British National Corpus (BNC), the Brown Corpus and the Susanne Corpus, which consists of 130,000 words of the Brown Corpus which have been analyzed syntactically. Written corpora can appear in many forms. These forms differ as much at the level of their structures and linguistic functions as at the level of their collection method.

– Verbal dictations: these are often texts read by office software users to gather digital texts in the form of data. Speakers vary in age range and it is necessary to record speakers of different genders to guarantee phonetic variation. Sometimes, geographical variations are also included, for example (in case of American English), New York English versus Midwest English.

– Spoken commands: this kind of corpus is made up of a collection of commands whose purpose is to control a machine such as a television or a robot. The structures of utterances used are often quite limited because short imperative sentences are naturally quite frequently used. Performance phenomena such as hesitation, self-correction or incompleteness are not very common.

– Human–machine dialogues: in this kind of corpus, we try to capture a spoken exchange or a written exchange between a human user and a computer. The diversity of linguistic phenomena that we are able to observe is quite limited. The main gaps come from the fact that machines are far from being as good as humans. Therefore, humans adapt to the level of the machine by simplifying their utterances [LUZ 95].

– Human–human dialogues mediated by machines: here, we have an exchange (spoken or written) between two different human users. The mediator role of the machine could quite simply involve transmitting written sequences or sound waves (often with some extent of loss in sound quality). Machines could also be more directly involved, especially in the case of translation systems. An example of such situation could be a speaker "A" who is speaking in French and this person who tries to reserve a hotel room in Tokyo by speaking to a Japanese agent (speaker B) who does not speak French.

– Multimodal dialogues: whether they are between a human and a machine or mediated by a machine, these dialogues have the ability to combine gestures and words. For example, in a drawing task, the user could ask the machine to move a blue square from one place to another. Put this square <pointing gesture towards the blue square> here <pointing gesture towards the desired location>.

1.2.2. *The historical point of view*

The period that a linguistic corpus represents can be considered as a criterion for distinguishing between corpora. There are corpora representing linguistic usage at a specific period in the history of a given language. The data covered by ancient texts often consist of a collection of literary texts and official texts (political speeches, archives of a state). In view of the fleeting nature of oral speech, it is virtually impossible to accurately identify all the sensitivities of a spoken language long ago.

1.2.3. *The language of corpora*

A corpus must be expressed in one or several languages. This leads us to need to distinguish between: monolingual corpora, multilingual corpora or parallel corpora.

Monolingual corpora are corpora whose content is formulated with the help of a single language. The majority of corpora that are available today are of this type. Thus, examples of corpora of this type are very common: the Brown Corpus and the Switchboard Corpus for written and spoken English, respectively, and the Frantext corpus, as well as the OTG corpus for written and spoken French, respectively.

Furthermore, parallel corpora include a collection of texts where versions of the text in several languages are connected to one another. These corpora can be represented as a graph or even a matrix of two dimensions n x m: where n is the number of texts (Tx) in the source language and m is the number of languages. News reports from press agencies such as Agence France-Presse (AFP) or Reuters are classic examples of sources of such corpora: each report is translated into several languages. Furthermore, several organizations and international companies such as the United Nations, the Canadian Parliament and Caterpillar have parallel corpora for various purposes. Some research laboratories have also collected this type of corpora, such as the European corpus CRATER by the University of Lancaster, which is a parallel corpus in English, French and Spanish. For a corpus to really be useful, fine alignments must be made at levels such as sentence or word. Thus, each sentence from text "T1" in language "L1" must be connected to a sentence in text "T2" in language "L2". An extract from a parallel corpus with aligned sentences is shown in Figure 1.1.

---

sub d = 22 ----------&
the location register should as a minimum contain the following information about a mobile station :
-----&
l'enregistreur de localisation doit contenir au moins les renseignements suivants sur une station mobile:
sub d = 386 ----------&
handover is the action of switching a call in progress from one cell to another (or radio channels in the same cell).
-----&
le transfert intercellulaire consiste à commuter une communication en cours d'une cellule (ou d'une voie radioélectrique à l'autre à l'intérieur de la même cellule).

Figure 1.1. *Extract from a parallel corpus [MCE 96]*

Note that a multitude of multilingual corpora exist which are not parallel corpora. For example, the corpus CALLFRIEND Collection is a corpus of telephone conversations available in 12 languages and three dialects, and the corpus CALLHOME is made up of telephone conversations available in six languages. In these two corpora, the dialogues, which are not identical from one language to another, are not connected in the same way as in the format presented above.

Parallel corpora are a fundamental source used to build and test machine translation software (see [KOE 05]). An important question to ask after having identified multilingual data is the alignment of the content of these data. To resolve such a fundamental problem to make use of multilingual corpora, a number of approaches have been proposed. Some approaches are based on the comparison of the length of sentences in terms of the number of characters they contain [GAL 93] and in terms of the number of words [BRO 91], while others adopt the criterion of vectorial distance between the segments of the corpora considered [FUN 94]. Furthermore, there are approaches which make use of lexical information to establish links between two aligned texts [CHE 93]. Other approaches combine the length of sentences with lexical information [MEL 99, MOO 02]. Note that the GIZA++ toolbox is particularly popular for aligning multilingual corpora.

1.2.4. *Thematic representativity*

This criterion affects written corpora which target the representativity of an entire language or at least a large proportion of this language. To achieve representativity at such a broad level, having a selection of texts coming from a variety of domains is essential. Three types of layouts can be cited:

– Balanced corpora: to guarantee thematic representativeness, texts are collected according to their topics, so as to ensure that each topic is represented equally.

– Pyramidal corpora: in these cases, corpora are constructed using large collections for topics considered central and small collections for topics considered less important.

– Opportunistic corpora: this kind of corpora is used in cases where there are not enough linguistic resources for a given language or for a given application. Therefore, it is indispensable to make the most of all available resources, even if they are not sufficient to guarantee the representativeness aimed for.

Note that guaranteeing the topic representativity of a corpus is often complicated. In most cases, texts look at several different topics at once and it is difficult (especially in the case of an automatic collection from a corpus, with the help of a web crawler, for example) to decide exactly what topic a given text covers. Moreover, as [DEW 98] underlines, there is no commonly accepted typology used for the classification of texts. Finally, it may be useful to mention that lexicography and online information research are among the areas of application which are the most sensitive to thematic representativeness.

1.2.5. *Age range of speakers*

The application or scientific domains often impose constraints regarding the age range of speakers. Certain corpora are only made up of linguistic productions uttered by adult speakers, such as air travel information system (ATIS), distributed by LDC. Certain corpora that will be used to research first language acquisition are made up of baby utterances. The most well-known example of this is the child language data exchange systems (CHILDES) corpus, collected and distributed at Carnegie Mellon University in the United States. Finally, corpora exist which cover the linguistic productions of adolescents, such as the spoken conversation corpora collected at the University of Southern Demark SDU as part of the European project NICE.

1.3. Who collects and distributes corpora?

The increasingly central role of corpora in the process of creating AI applications has led to the emergence of numerous organizations and projects with a mission to create, transcribe, annotate and distribute corpora.

1.3.1. *The Gutenberg project[1]*

This is a multilingual library which distributes approximately 45,000 free books. This project makes an extensive choice of books available to Internet users, both at the linguistic level and at the level of topics available, since it distributes literary works, scientific works, historical works, etc. Nevertheless, since it is not specifically designed to be used as a corpus, the works distributed in this project need some preprocessing to make them usable as a corpus.

1.3.2. *The linguistic data consortium*

Founded in 1992 and based at the University of Pennsylvania in the United States, this research and development center is financed primarily by the National Science Foundation (NSF). Its main activities consist of collecting, distributing and annotating linguistic resources which correspond to the needs of research centers and American companies which work in the field of language technology. The linguistic data consortium (LDC) owns an extensive catalog of written and spoken corpora which covers a fairly large number of different languages.

1.3.3. *European language resource agency*

This is a European level centralized not-for-profit organization. Since its creation in 1995, the European language resource agency (ELRA[2]) has been collecting, distributing and validating spoken, written and terminological linguistic resources, as well as software tools. Although it is based in the European city of Paris, this organization does not only look at European languages. Indeed, many corpora of non-European languages, including Arabic, feature in its catalog. Among its scientific activities, the ELRA organizes a biannual conference: language resources and evaluation conference (LREC).

---

1 https://www.gutenberg.org/.
2 http://www.elra.info/en/.

1.3.4. *Open language archives community*

Open language archives community (OLAC[3]) is a consortium of institutions and individuals which is creating a virtual library of linguistic resources on a global scale and is developing a consensus on best practices for the digital archiving of linguistic resources by creating a network of storing services for these resources.

1.3.5. *Miscellaneous*

Given the considerable costs of a quality corpus and the lucrative character of most existing organizations, it is often difficult for researchers who do not have a sufficient budget to get hold of corpora that they need for their studies. Moreover, many manufacturers and research laboratories jealously keep back the linguistic resources they own, even after the projects for which the corpora were collected have finished.

To confront this problem of accessibility, many centers and laboratories have begun to adopt a logic that is similar to that of free software. Laboratories such as CLIPS-IMAG and Valoria have, for example, taken the initiative of collecting and distributing two corpora of oral dialogues for free. These corpora include the Grenoble Tourism Office corpus and the Massy School corpus[4] [ANT 02]. In the United States, there are examples such as the Trains Corpus collected by the University of Rochester, whose transcriptions have been made readily available to the community [HEE 95]. In addition, the *ngrams of the Google books*[5] is a corpus which is used more and more for various purposes.

1.4. The lifecycle of a corpus

As an artificial object, corpora can only very rarely exist in the natural world. Corpora collection often requires important resources. From this point of view, in some ways, the lifecycle of a corpus resembles the lifecycle of a piece of software. To get a closer look at the lifecycle of a corpus, let us examine the flowchart shown in Figure 1.2. As we can see that there are four

---

3 http://www.language-archives.org/.
4 http://www.info.univ-tours.fr/~antoine/parole_publique/Massy/index.html.
5 https://books.google.com/ngrams.

main steps involved in this process: preparation/planning, acquisition and preparation of the data, use of the data and evaluation of the data. It is a cyclical process and certain steps are repeated to deal with a lack of linguistic representativeness (often diachronic, geographical or empirical in nature) to improve the results of an NLP module.

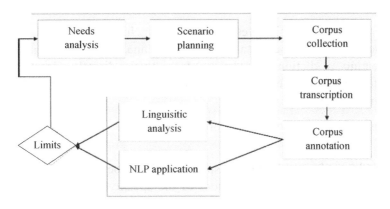

Figure 1.2. *Lifecycle of a corpus*

Three main steps stand out within a lifecycle:

– The preparatory step: this is about the work carried out before the corpus collection. In this step, key questions must be answered, such as: Why do we need a corpus? What properties should such a corpus have? How can we collect this corpus?

– The collection and the annotation of the corpus: this step covers the work necessary to construct the corpus in such a way that the objectives fixed in the preceding step can be reached.

– The use of the corpus: this step is about the statistical analysis and/or the linguistic analysis of the contents of the corpus. This step can bring some insights into the studied linguistic subject. For example, you can try to calculate the number of syntactic constructions by knowing the thematic context or the type of text (medical text, journalistic text, etc.). Moreover, the corpus can be used to construct NLP modules.

As we shall see later on, the lifecycle of a spoken corpus is distinguished by an additional step which is the transcription of spoken utterances. Moreover, given their situation in a specific spatio-temporal context, dialogue corpora (both written and spoken) require the definition of scenarios to ensure a minimal level of representativeness of the dialogue domain.

1.4.1. *Needs analysis*

Examples of the objectives of a corpus include analyzing varieties of syntactic styles, constructing a morphological analyzer for a given language and creating a dictionary. The needs analysis directly affects all the parameters which define the type of a corpus. Among others, this allows the following to be decided:

– Basic choices: whether the corpus is spoken or written, the languages, etc.

– Speakers: the age range of speakers, their socioeconomic status, the number of speakers used, the gender of the speakers (percentage of males and females).

– Size of the corpus: when we have to collect a corpus to make a dictionary for the Arabic language, for example, we need to use a very broad corpus to make sure that all the linguistic registers and all socioeconomic factors have been taken into account.

– Thematic structure of the corpus: pyramidal, balanced, etc.

1.4.2. *Design of scenarios to collect data for the corpus*

After having specified the collection objectives, the linguists must describe how the corpus is to be collected. This must happen according to the objectives specified in the preceding step. Note that the scenarios used for collection involve both spoken and written conversation corpora and that one scenario can sometimes be adapted to several collection methods.

1.4.3. *Collection of the corpus*

As we have already seen, a corpus is a collection of texts that is specifically selected to satisfy a number of predetermined constraints. The simplest way of collecting a corpus is to use real existing data. As far as spoken data is concerned, the broadcast news is probably the most well-known example. It consists of a televised news program accompanied by a written transcription. For written data, the Internet is incontestably the most abundant source. This is also reflected by the diversity of the linguistic forms and registers available online such as classical literature, informal chat and discussion forums.

Collection is carried out using a web crawler, which collects information automatically according to predefined thematic and linguistic criteria. Creating a list of documents can be done in two different ways. One way is to do this using a search engine: in this case, the crawler uses a number of keywords which it successively submits to one or several search engines. The URLs collected from the search results are added to the list of documents to be analyzed. The search engine plays the role of a topic filter here since only pages corresponding to the query topic are obtained. The other way is to obtain the list of documents using a list of URLs. This list can be initialized right at the beginning with a collection of links generated manually. Next, new URLs are extracted from the pages visited and are used to expand the list of URLs to be visited. This allows an exploration of the document space using a *Breadth First Search* approach. Note that crawlers must respect the rules of ethics which involve consuming the minimum amount of resources from the server from which the data are extracted. Often, crawlers are equipped with a language detection algorithm. An algorithm like this is able to classify the documents according to the language they are written with. Thus, the language and the theme of the text are, in general, the main selection criteria for a page to be included in a database. NLP specialists have made use of this source of information in the development of several types of applications, including speech recognition software and the POS tagging (see [VAU 00]).

In some cases, linguists use computer programs to generate sentences which correspond mainly to syntactic criteria. Among the most well-adapted tools is definite clause grammar (DCG), developed using the PROLOG language (logic programming). Due to the limitations of current automatic generation systems, it is often considered to be costly to constrain the syntactic grammars used for this kind of objective using semantic criteria. Thus, such corpora are of no interest to linguistic research. Often, they are used to train speech recognition modules (in particular, statistical language models). The main aim of this method is to obtain a number of syntactically acceptable texts with a minimal amount of time and effort.

To collect linguistic data that conform to specific criteria, it is possible to create a description of the system's task, which can then be used as a support for the generation of data. For example, at the University of Aleppo, in the framework of the construction of the prototype of our system AraTis (airline reservation system in Arabic), we carried out data collection of this type, since at the beginning of the project, no linguistic data of this type were freely available. The advantage of this method is that no special preparations are required. The only requirements to collect data of a reasonable quality and quantity include having a clear description of the system's task and getting a sufficient number of speakers. The number of speakers varies naturally from one application to another when collecting data of reasonable quality and quantity. Previous works have shown that the task as well as the physical context influence the linguistic behavior of speakers [LUZ 95]. This limits the possibilities of using such data for the rapid development of prototypes since the statistical representativeness of the phenomena is not guaranteed. The Wizard of Oz method is often used to address these shortcomings.

To develop a human–machine dialogue system of any type, we need to model several sources of knowledge at different levels. This includes linguistic and metalinguistic knowledge, which involve a considerable number of factors which directly influence how conversations progress. Besides, this knowledge includes information about the speaker, the speaker's way of speaking, the speaker's linguistic level (whether they are native or foreign). In addition, this knowledge includes information about the conversation topic, how certain operations are carried out and knowledge of the physical context, i.e. where the dialogue takes place (e.g. at a train station, at an airport or at the workplace, etc.).

To take into consideration all the knowledge that we have just outlined and to simulate the behavior of speakers when faced with a real system before its creation, researchers use the Wizard of Oz method. The idea of this method is to put the participant in a context which makes him think that he is interacting with an intelligent computer program, but in reality, he is interacting with a fellow human who is simulating the reactions of the machine. This is shown in the diagram outlined in Figure 1.3.

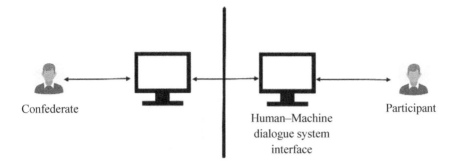

Confederate   Human–Machine   Participant
                               dialogue system
                               interface

Figure 1.3. *Data collection system using the Wizard of Oz method*

The main advantage of the Wizard of Oz method is that it comes close to real utilization conditions and, therefore, the data produced is of a better quality, both linguistically and in terms of the knowledge linked to the applied usage of such data. However, in some cases, the cost and the tools necessary for collection can exceed the financial means of most laboratories. For some projects, such as those involving a dialog with an embedded system in a car or an airplane, we need to use simulators for these machines which makes the project extremely expensive. Therefore, only large specialized companies are able to carry out collection using tools of this kind (see [GEU 02] for an example of a collection for a corpus using a car simulator).

Manually collected corpora, or sometimes corpora collected using the Wizard of Oz method, are often used to develop a preliminary version of a system or a prototype. This prototype can be used to collect better quality data, which, in turn, can be used to improve the performance of the prototype itself (see Figure 1.4). For example, we can cite the Halpin system, which was developed within the laboratory CLIPS-IMAG [ROU 00]. This system of human–machine dialogue that can be used to research bibliographic references in the IMAG media library was put online to collect usage data. This data is used later to improve the quality of the system. Successive versions of the system were released, and at each iteration, the quality of the system improved and consequently the quality of the data collected was also improved.

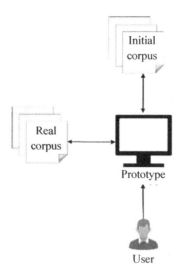

Figure 1.4. *Diagram of a corpus data collection system using a prototype*

Therefore, this is an incremental process that can continue as long as the system is in use. In this way, it is possible to take into account the potential evolutions of the linguistic and interactive behavior of users. Using prototypes for collection is a very good way of obtaining real data easily. On the other hand, this method requires a lot of resources to annotate the large quantities of data which are obtained in this way. Furthermore, since the prototype is made readily available to users, some users occasionally take the system for a game and, therefore, do not produce utterances that correspond to the purpose of the system. To filter out such utterances, extra effort is required.

1.4.4. *Transcription*

Transcription involves producing a written version of a recording obtained using one of the different collection methods. A professional transcription must be carried out rigorously and three fundamental principles must be respected [EDW 93]:

– categories must be discriminating, exhaustive and contrastive;

– transcriptions must be easy to read;

– transcriptions must be systematic and predictable to make the automatic processing of data possible.

Before beginning the transcription, the type of transcription must be decided, in order to know whether an orthographic, phonetic or prosodic (or a combination) transcription is required. If a combination is required, the transcriptions must be aligned. An agreement must be reached as much for the language in question as for foreign words regarding problematic spellings, which can be quite common in transcriptions, e.g. alternative spellings for Kuwait in the French language (*Kuweit*, *Koweït* and *Koweit)*. This is necessary to guarantee the homogeneity of transcriptions. In the same way, it is important to plan to take into account non-verbal phenomena present in the speech signal when they are produced by speakers, e.g. clicks, coughs, hesitations and long or short pauses. Short pauses are typically between 0.2 and 0.5 seconds while long pauses are those whose length exceeds 0.5 seconds. Equally, it is possible to consider sound phenomena linked to the environment where the conversation is being recorded such as objects falling, parallel conversations and the noise of cars or airplanes.

Often, the software used in transcription offers an open list in which the user can insert labels to be used for certain phenomena. Let us look at Figure 1.5, which gives us an example of the transcription of a radio sequence produced using the transcription software Transcriber[6], which is distributed under the general public license (GPL). In this example, each speaker's contribution begins with the name given to them by the transcribers. In the case of our example, we have two speakers: Simon Tivolle and Patricia Martin. The two first speaking turns are marked #1 and #2, respectively, which signifies that the two turns are happening in parallel. The labels [laugh] and [i] indicate, respectively, that a laugh and an inhalation occurred at that moment (by their presence in the sequence). Finally, the labels [laugh-] and [-laugh] show that the sequence between them is produced in parallel with a laugh.

Simon Tivolle : #1 yeah. #
Patricia Martin : #2 sure ? #
Simon Tivolle : really? [laugh] no. joke, Patricia's joke. [i] France-Inter, [laugh-] it's 7 o'clock [-laugh].
Patricia Martin : the news, Simon Tivolle:
Simon Tivolle : [i] hello! Tuesday April 28th. The national consultation on the national high school: [i] a huge debate today and tomorrow in Lyon to learn about

Figure 1.5. *Transcription example using the software Transcriber*

---

6 http://www.etca.fr/CTA/gip/Projets/Transcriber/.

Finally, note that the process of transcribing large amounts of data requires the implementation of a hierarchical cooperation process between several linguists to verify the transcriptions more than once and, therefore, ensure that the quality required is achieved.

1.4.5. *Corpus annotation*

Annotation is the process which involves enhancing the text with linguistic information or sometimes general information that describes the contents of the corpus. In other words, annotation involves adding value to the corpus, since it improves its quality and, therefore, opens up the ways in which the corpus can be used (see [PAL 10] and [PUS 12] for a general introduction to this). Annotation typically corresponds to the levels of linguistic structure: morphology, syntax, semantics, etc. The annotation of a corpus with non-linguistic information is also possible. Annotation can be carried out manually when appropriate, but very often, NLP tools are used to carry out annotation automatically. In this case, a checking and error correction phase is indispensable. A good annotation must always be well documented to guide users. It must be as neutral as possible regarding theoretical controversies to maximize the scope of its usage.

The first step in the annotation process is the raw corpus made up of *tokenized* but unannotated texts which are cleaned to remove special characters, if necessary. Sometimes, depending on the type of text, titles and paragraphs are marked.

Texts annotated with parts of speech are one of the most commonly used corpora. This kind of corpora are annotated using POS tags. This corpora is mainly used to build and test parts of speech taggers or to test syntactic parsers. An example of a fragment of text annotated using parts of speech is shown in Figure 1.6.

a.	SpeakerB3/SYM./.
b.	Well/UH what/WP do/VBP you/PRP think/VB about/IN the/DT idea/NN of/IN ,/, uh/UH ,/, kids/NNS having/VBG to/TO do/VB public/JJ service/NN work/NN for/IN a/DT year/NN?/.

Figure 1.6. *Segment of a corpus analyzed using parts of speech*

As we will see in the chapter of syntax, statistical parsing has made significant progress, especially in terms of robustness and the resolution of ambiguity, thanks to the availability of syntactically annotated corpora. In practice, the realization of these parsers requires syntactically parsed data which are commonly named treebanks. A grammar based parser is usually used to annotate the corpus syntactically. Next, linguists begin to review the annotated corpus to be able to correct the inevitable errors introduced by the parser.

To be widely usable, it is important these corpora are independent of existing syntactic theories. However, there are two main schools of thought within the linguistics community, namely the structuralism and the functionalism, mirroring the famous schism in syntax.

The structuralism focuses on noun phrases, verb phrases, etc. The *Penn Treebank* by the University of Pennsylvania is the most popular example of this type of treebanks [MAR 94]. It is made up of a syntactically annotated collection of sentences from the Brown Corpus and the Switchboard Corpus. An example of a simple sentence from the Penn Treebank is shown in Figure 1.7.

```
( (CODE SpeakerB1 .))
( (INTJ Okay . E_S))
( (CODE SpeakerA2 .))
( (INTJ Okay . E_S))
( (CODE SpeakerB3 .))
( (SBARQ (INTJ Well)
            (WHNP-1 what)
            (SQ do
              (NP-SBJ you)
              (VP think
                        (NP *T*-1)
                        (PP about
                          (NP (NP the idea)
                              (PP of

                                 ,
                                (INTJ uh)

                                 ,
                                (S-NOM (NP-SBJ-2 kids)
                                       (VP having
                                                 (S (NP-SBJ *-2)
                                                    (VP to
                                                      (VP do
                                                                 (NP public
service work))))
                                                      (PP-TMP for
                                                              (NP a
year)))))))))
```

Figure 1.7. *Extract from the Penn Treebank*

As we can see in Figure 1.7, the sentences are labeled in the style of the programming language Lisp rather than XML.

A tree corpus for French was also constructed at the Formal Linguistics Lab (LLF) at Denis-Diderot University in Paris [ABE 03]. Made up of about 22,000 sentences and 870,000 words, this corpus was created by extracting sections of the daily newspaper *Le Monde* that appeared in 1990, 1992 and 1993. The corpus covers texts written by a number of authors on varying subjects from economics to literature and politics, etc. In contrast to the Penn Treebank, this corpus used a format based on XML, as shown in Figure 1.8. It has been distributed freely since 2001.

```
<SENT nb="7">
<PP fct="MOD"> Parmi
            <NP> les candidats
            <PP>à
                        <NP> la commission exécutive
                                    <PP> de <NP> La CGT </NP>
                        </PP>
            </NP>
            </PP>
</NP>
</PP> ,
<VN fct="SUJ"> on compte </VN>
<NP fct="OBJ"> quarante---quatre nouveaux---venus </NP>.</SENT>
```

Figure 1.8. *Extract from a tree corpus for French*

Functional annotation uses a radically different approach and focuses on syntactic relationships and dependencies between words. This is the case in the *Prague Dependency Treebank* and the *English Dependency Treebank* [HAJ 98]. In fact, [XIA 01] showed that it is not possible to convert a dependency tree corpus into a corpus annotated using the structural approach such as the Penn Treebank because the functional approach treats the subject and object equally regarding their attachment to the verb.

There are corpora which are semantically annotated. In contrast to syntactic annotation, semantic annotation approaches are quite diverse and fulfill a number of purposes. Some annotations cover semantic relationships between constituents in the sentence, e.g. the Proposition Bank [PAL 05].

Annotated at the University of Lancaster in the UK, the clinical text corpus, CLEF, is another example of a corpus of this type. Among the semantic relationships considered by this corpus, there is the *has_target* which compares an intervention or an investigation using the part of the corpus in question. It is, therefore, a predicate (relationship) which takes two arguments. The first argument is *investigation* or *intervention* and the second is *zone*.

```
This patient has had a [arg2 lymph node]
[arg1 biopsy]
… he does need a [arg2 groin]
[arg1 dissection]
```

Figure 1.9. *Semantic annotation with a has_target relationship*

In the first sentence of the example shown in Figure 1.9, the predicate is *has had*, the intervention is *biopsy* and the zone of intervention is *lymph node*. The corpus GENIA is another semantically annotated medical corpus. It is a corpus which will be used to facilitate the extraction of knowledge based on genetic data [KIM 03]. Another form of annotation involves using temporal expressions such as those in the TimeBank [PUS 03].

There are also corpora which are annotated with discursive relationships, for example the RST Corpus, which is made up of 385 articles extracted from the Penn Treebank[7]. It is hierarchically annotated according to the rhetorical structure theory (RST) by [MAN 88]. The main task involved in annotation consists of identifying the elementary discursive units (EDUs). The discursive tree corpus Discourse Treebank from the University of Pennsylvania adopted an approach which was more centered on discursive connectors and their arguments [MIL 04]. It is probably useful to mention the annotation of co-referential relationships in the corpus by [POE 04] and the corpus of opinions [WIE 05].

Finally, it is probably worth mentioning some existing annotation tools. EXMARaLDA[8] is a German multi-level annotation tool which is entirely based on XML language. Specially adapted to discursive annotation, it contains a data annotation tool, a corpus manager which combines annotated files and adds the metadata. Developed at the Universidad

---

7 http://www.isi.edu/~marcu/discourse/Corpora.html.
8 http://www.exmaralda.org/.

Autónoma de Madrid, the UAM Corpus Tool[9] is another annotation tool, designed to be user-friendly to make annotation easier for linguists whose programming skills are limited [O'DO 08]. It is distributed with a number of NLP and research tools for English. The Brat Rapid Annotation Tool[10] by MIT is another example of an annotation tool. With a web interface, it is particularly adapted to collaborative annotation projects. It was used in projects about entity and event detection and extraction, as well as in projects about shallow parsing, etc. Other tools whose aims are more specific should also be mentioned. For example, CLaRK[11] for the annotation of syntactic information, NITE[12] for multimodal annotations and MMAX2[13] for anaphor annotation.

1.4.6. *Corpus documentation*

The aim of the documentation is to make corpora accessible to the community. Typically, three files are used to document corpora. Firstly, there is the initial file which is commonly called *readme*. This file contains information about the rights of authors, the version of the corpus, information about the corpus documentation (the other files) as well as summary information about the corpus: the size, the number of speakers, structure, etc. This is followed by the documentation file which includes a detailed description of all aspects of the corpus. Among other things, this includes the recruitment criteria for participants (e.g. age range, socioeconomic status, etc.), the annotation procedure, the format used, the software used, the recording and metadata. Finally, specific documents are put together to cover specific aspects of the corpus such as the history of the corpus, internal publications on the corpus in the form of technical reports, etc.

1.4.7. *Statistical analysis of data*

The statistical analysis of data involves looking at the frequency, the mean and the median of particular phenomena such as the frequency of a

---

9 http://www.wagsoft.com/CorpusTool/.
10 http://brat.nlplab.org.
11 http://www.bultreebank.org/clark/index.html.
12 http://www.ltg.ed.ac.uk/NITE/.
13 http://mmax2.net.

certain word or word category, a syntactic structure, an opinion, or another discursive phenomenon. It is possible to carry out the description of a given corpus or to compare these phenomena in two or several corpora.

1.4.8. *The use of corpora in NLP*

The way in which corpora are used to construct an NLP module depends on the approach used for processing. Rule-based approaches do not require specific annotations, since it is the responsibility of the human developer to extract the knowledge from the corpus as he or she sees fit. In contrast, learning-based approaches require annotated data to guide the process of information extraction and processing. The degree of granularity of the annotation required varies considerably according to the applicative aim of the module, as well as the algorithm and the approach that it adopts, such as whether it involves supervised or unsupervised learning, neural networks, statistical algorithms, and automatic grammar induction algorithms.

1.5. Examples of existing corpora

1.5.1. *American National Corpus*

This non-free corpus has the objective of collecting a million words from transcribed spoken data, as well as a collection of written texts whose size is approximately ten million words. The American National Corpus (ANC) team is made up of people in industry, as well as academic teams. This corpus includes important sections that are annotated with POS tags and is distributed using the XML coding standard format (XCES).

1.5.2. *Oxford English Corpus*

The Oxford English Corpus (OEC) is a collection of English texts which was used to support the creation of the *Oxford English Dictionary*, published by Oxford University Press. Containing more than two billion words, it is the largest corpus of its kind in the world. The texts which make up this corpus are extremely varied. Literary texts, specialized newspapers, daily newspapers, weekly newspapers, websites, and extracts of forums, among other types of texts, make up the main source of this corpus.

The OEC is annotated with XML and is often analyzed with the software Sketch Engine. Each document of the OEC is accompanied with the following metadata:

– title;

– author (if known);

– type of author (if known);

– dialect (British English, US English, etc.);

– source (website);

– date of the document (if known);

– date it was added to the corpus;

– field and sub-field;

– document statistics (number of tokens, sentences, etc.)

1.5.3. *The Grenoble Tourism Office Corpus*

Recorded by the laboratory CLIPS-IMAG in the Grenoble Tourism Office, this is a collection of task-oriented human–human spoken dialogues which come from the applied setting of tourist information [ANT 02]. The collection of data is carried out in real conditions following a semi-blind method: it involves an interaction between a member of the tourism office team and members of the public who are visiting the town. The real life conditions for recording meant that some sound quality was lost. The recordings were carried out on two different paths using a digital audio tape (DAT) recorder. In this way, two audio files in .wav format were obtained per conversation. In total, seven hours of recording were obtained. This corpus was initially limited to being distributed to members of the ARC. Today, it is distributed in two formats, the transcribed corpus can be downloaded directly from a web page associated with the project PAROLE PUBLIQUE[14] and the complete corpus (transcription and audio files), due to the size of the audio files, is distributed on CDs by post.

---

14 http://www-valoria.univ-ubs.fr/antoine/parole_publique.

The Sphere of Speech

2.1. Linguistic studies of speech

The scientific study of speech, found at the intersection of a number of disciplines such as Physiology, Electronics, Psychology and Linguistics, is a very rich field. It forms the basis of a large number of practical applications, including speech pathology, language teaching and speech processing (SP), which is the focus of this chapter. Two key disciplines are at the heart of studies in this area: phonetics and phonology.

2.1.1. *Phonetics*

Phonetics is a field that deals with the scientific study of speech sounds. It provides methods for the description, classification and transcription of speech sounds [CRY 91]. As a branch of the natural sciences, it examines speech sounds from physical, physiological and psychological points of view. Phonetics involves making parallels between sounds in human language and looking at the physiological, physical and psychological constraints which have shaped these sounds. For a general introduction to phonetics, the reader can refer to a number of excellent works, including those of [LÉO 05, LAD 01, HAR 99, CAR 74].

Historically, the interest in describing the sounds of speech goes back to antiquity. It is widely thought that the Phoenicians developed the first system of phonetic transcription or alphabet on the east coast of the Mediterranean between 1700 and 1500 BC. The traces of this alphabet are still visible in some alphabets that exist today. As for the first actual phonetic descriptions, it is thought that they began in around 500 BC with the grammar of Sanskrit written by Panini, which contains a remarkable classification of the sounds of this language. In the Middle Ages, Arabic linguists carried out advanced research on descriptions of the sounds of Arabic, notably using a system of phonetic contrasts. Some of these works had a prescriptive objective, as the aim was to conserve the original phonetic form of the Quran's message, eliminating changes which would take place throughout the ages [CRY 71]. Others, such as Avicenna's work, had the aim of understanding the physiological basics of the process of speech production and, therefore, were more scientific in nature (see [KAD 04, WER 84] for a review of these works). In Europe, the interest in phonetics began to develop at the beginning of the 18th century, with the works of Joshua Steele, among others. It was in the 19th century, with Thomas Edison's invention of the phonograph, that phonetics was able to make a large leap in progress. Thanks to this device, phoneticians were able to record spoken sequences and could describe their properties (by slowing down or speeding up the recordings). Among the works written during this period, we can cite those of Ludimar Hermann on the production of vowels.

Since it is much more practical to work with written forms, phoneticians developed a number of transcription systems for transcribing speech. In contrast to an ordinary alphabet, a transcription system is different because there is a direct and unique correspondence between a sound and its grapheme. Each individual sound corresponds to a single grapheme in the system of transcription and each grapheme is associated with one sound. Furthermore, a transcription system must be universal to be used to transcribe all the languages in the world. Among the systems of transcription, the *International Phonetic Alphabet* (IPA) is the best-known. It is a transcription system developed towards the end of the 19th century by the IPA. To examine the purpose of such an alphabet, let us examine Table 2.1, which presents some graphical forms in French for the sound [o] and some examples of transcription of the grapheme "o" in English.

Word	Transcription
rose	[Rɔz]
chaud	[ʃo]
Beau	[bo]
show	[ʃoʊ]
bond	[bɑnd]
drone	[drəʊn]
ubiquitous	[jʊ'bɪkwɪt̬əs]

Table 2.1. *Examples of IPA transcriptions from French and English*

The differences in grammatical forms are due to several reasons which can be morphological or even etymological. In spite of their apparent banality, the sounds of a language are far from being simple entities to study. Sounds can be studied from several points of view according to the specific location in the communication system (see Figure 2.1).

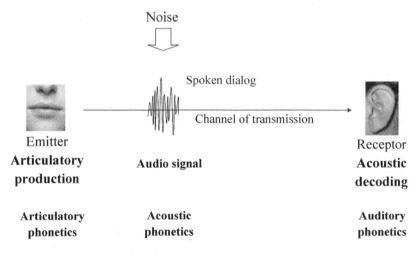

Figure 2.1. *Communication system*

This shows how the three main branches of phonetics are distinct: articulatory phonetics, which focuses on the production of sounds by the emitter; acoustic phonetics, which is located at the level of the transmission channel; and auditory phonetics, which focuses on the way in which the sounds are perceived by the receptor.

2.1.1.1. *Articulatory phonetics*

Articulatory phonetics takes the point of view of the emitter and considers the way in which sounds are produced by the speech organs such as the pharynx, the tongue and the lips. In other words, the physiology of the production of sounds is the true focus of this branch of phonetics.

To fully understand the process of the production of speech sounds, we can compare it to a well-known artificial process – the production of music by instruments. In both cases, we need a source which can be modified by a sound box of sorts. We are talking about a complex process involving a number of organs (the organs that make up the vocal tract) whose roles need to be coordinated efficiently. To explain this in a systematic way, these organs can be grouped into three functional components: the subglottic system, the phonatory system and the supraglottic (supralaryngeal) system (see Figure 2.2(a) for a global view of the speech apparatus and Figure 2.2(b) for a lateral view of the phonatory and supraglottic systems). Moreover, Broca's area, situated in the frontal lobe of the cerebral cortex, can be considered as a major component of the vocal apparatus, because of the leading role that it plays in the coordination of this process. For more information on the process of speech production, see [MAR 07].

The *subglottic system* has the role of creating an air current following the deflation of the lungs after the diaphragm relaxes. Air freed in this way then moves across the trachea before arriving at the larynx. The main component of the production of sound, the *larynx*, is a cartilaginous solid structure which is situated above the trachea and below the middle part of the pharynx. Another specificity of the larynx is that it contains the two *vocal folds* which are two muscular bands of about 1 cm in length and 3 mm in width. During phonation, these folds come close to one another and begin to vibrate against each other. This vibration gives the sound its basic quality and is called the *fundamental frequency*. Think of vibration as happening in two modes: a heavy mode which produces deep sounds and a light mode for high-pitched sounds.

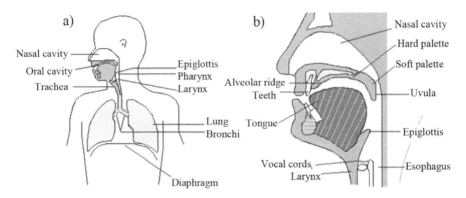

Figure 2.2. *Speech organs*

The articulation system, located above the larynx, is made up of three cavities which play an important role in the formation of sounds: the pharynx, the nasal cavity and the oral cavity. The *pharynx* is a cavity which resembles a tube whose main role is to connect the larynx with the nasal cavities and the respiratory tract as well as the mouth and the esophagus (digestive tract). The shape of the pharyngeal cavity determines the fundamental qualities which give a unique timbre to the voice of each and every one of us. With a fixed size and shape, the *nasal cavity* plays the role of the sound box, filtering the sound and producing resonances. In order for the air to pass through this cavity, the soft palate (velum) must be lowered (see Figure 2.3). If the lips are closed, the entire mass of air being released is directed through the nasal cavity, producing consonants like *m*. Likewise, the opening of the lips accompanied by the lowering of the soft palate produces nasal vowels such as the sound [ɑ̃] in French, for example in the word *grand* [gʀɑ̃].

The *oral cavity* is the most important of the three cavities, because of its total volume and also because of the number of organs which are found within it. It is made up of the palate, the tongue, the lips and the lower jaw.

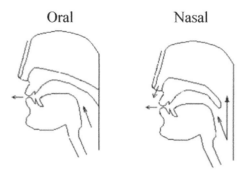

Figure 2.3. *Position of the soft palate during the production of French vowels*

The palate, the upper part of the oral cavity, is separated into two parts: the hard palate and the soft palate. The *hard palate*, located in the bony frontal part, is immobile and its role is to support the tongue during its movements, therefore allowing for partial occlusions to be made. The *soft palate* or *velum* is situated further back.

The tongue is a complex organ made up of a collection of muscles which are responsible for its large range of mobility. It is divided into three main parts: the apex, the body and the root (see Figure 2.4).

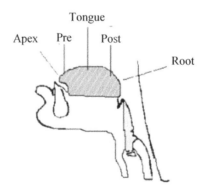

Figure 2.4. *Parts and aperture of the tongue*

The distance between the body of the tongue and the palate or the degree of aperture allows the quantity of air which passes through the mouth to be controlled (Figure 2.5).

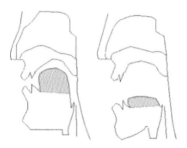

Figure 2.5. *Degree of aperture*

The *lips*, just like the tongue, are highly mobile and this allows them to assume many different shapes. Thus, when they come together, they create closure, for example when producing [m] or in the closure phase of [p]. On the other hand, when they are slightly open and round, they allow the production of vowels such as [o] and [u]. Likewise, when they are open and spread, they can produce vowels such as [i] and with a neutral position, they can produce [ə]. Finally, when the lower lip makes contact with the incisors, the sound [f] can be produced.

Ultimately, the lower *jaw* plays a moderating role by increasing or decreasing the size of the oral cavity.

2.1.1.2. *Acoustic phonetics*

Acoustic phonetics is a branch of phonetics which considers the physical properties of sound waves. Its focus is on what happens between the emitter and the receptor (see [MAR 08] for an introduction).

Let us begin by looking at the mechanism of production of sound waves. The movements of speech organs transmit a certain amount of energy to the air molecules surrounding them, creating a pattern of zones of high and low pressure, and we call this a sound wave. These waves propagate through the air (or, in some cases, through water), transmitting energy to neighboring molecules. To simplify the explanation of this process, let us look, for example, at the vibration of a tuning fork which causes the displacement of air molecules both forwards and backwards, as shown in Figure 2.6.

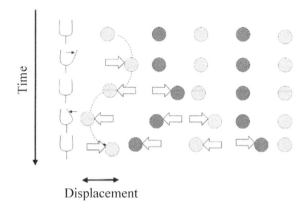

Displacement

Figure 2.6. *Displacement of air molecules by the vibrations of a tuning fork*

As minimal units of communication, sound waves must be able to be perceptibly distinguishable from one another by virtue of their fundamental physical properties. These include frequency and amplitude differences which cause perceptual variations in tone (pitch) and volume (loudness), respectively (see Figure 2.7).

The *frequency* of a signal is measured in cycles per second or Hertz, while *amplitude*, also sometimes called *intensity*, is a measure of the energy transmitted by a wave according to a unit called the *decibel*. Note that a tuning fork does not vibrate infinitely but even if its amplitude lowers little by little, its frequency remains stable. This means that frequency and amplitude are completely independent from one another. In other words, we can have two different signals that have the same frequency but different amplitudes, or two signals that have the same amplitude but different frequencies.

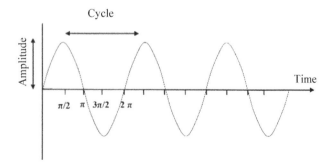

Figure 2.7. *Frequency and amplitude of a simple wave*

Note that the wave shown in Figure 2.7 presents a certain regularity because it does not change shape as time progresses: the distance between the peaks is always fixed. This kind of wave is called a repetitive wave or a periodic wave. The waves of vowel sounds are waves of this type. However, another kind of wave does not produce this kind of regularity and is logically called a non-repetitive or aperiodic wave. This kind of wave corresponds to plosive sounds such as [p] or [b] or to clicks (see Figure 2.8).

Figure 2.8. *An aperiodic wave*

It is also useful to distinguish between simple waves and complex waves. Complex waves are the result of combining multiple simple waves. Consequently, it is possible to analyze a complex wave by looking at its simple components. As illustrated in Figure 2.9, the presentation of a complex signal (a) in the form of simple waves (b) does not allow us to show all the properties of its components. Therefore, it is necessary to use a more sophisticated instrument such as a spectrograph to do this.

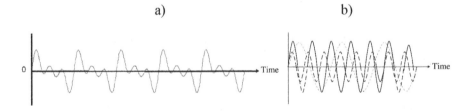

Figure 2.9. *Analysis of a complex wave*

Before presenting the spectrograph, we shall consider the term *resonance*. Resonance is the transmission of a vibration from one body to another, namely the resonator. Note that each body has its own natural frequency that it reacts to with the most ease. Thus, a tuning fork that vibrates at 200 Hz will cause the vibration of another tuning fork if it has a natural neighboring frequency. A simplified way of seeing the spectrograph is to consider it as a collection of tuning forks, each with a different natural frequency. When these tuning forks find themselves next to a complex wave, some, whose natural frequency corresponds to the frequency of a simple component of this wave, will resonate, making it possible to carry out an analysis of this wave (see Figure 2.10).

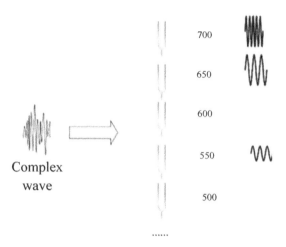

Figure 2.10. *A collection of tuning forks plays the role of a spectrograph*

In Figure 2.10, we have a complex wave made up of three components which will make three tuning forks of the same frequency vibrate. These frequencies are 550 Hz. The information that is produced by a spectrograph is called a *spectrogram*. A spectrogram is a multidimensional representation of the sounds obtained with the fast Fourier transform. As shown in Figure 2.11, the vertical axis represents frequency, whereas the horizontal axis represents time. The dark zones in the spectrogram give information about concentrations of acoustic energy and these are called *formants*.

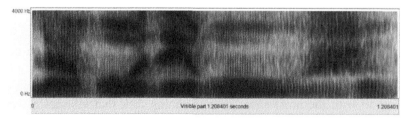

Figure 2.11. *Spectrogram of a French speaker saying "la rose est rouge" generated using the Praat software*

Formants play a fundamental role in the analysis of vowels where they are visible very clearly as vertical bars. To distinguish a vowel, it is normally sufficient to look at the first three formants. Table 2.2 presents the average values of the three first formants of the French vowels [a], [i] and [u].

Formant	[a]	[i]	[u]
F1	750	200	200
F2	1200	2200	700
F3	2600	3200	2200

Table 2.2. *The three first formants of the vowels [a], [i] and [u]*

In general, F1 correlates with the size of the pharyngeal cavity. This explains why the F1 value of the vowel [a], which is an open vowel, is the highest. Likewise, the values of the first formant of vowels [i] and [u] are very low, because they are close vowels. F2 is generally associated with the shape and size of the oral cavity. In the case of the vowel [i], where the oral cavity is reduced in size, this causes a very high F2 level. In the case of the vowel [a], the openness and the large amount of space within the oral cavity causes a lower F2 value. When it comes to F3, it is possible to distinguish between the French vowels [i] and [y], where the lips play an important role ([y] being a front rounded vowel that is not used in English). When professional singers perform, the lowering of the larynx and the raising of the tongue causes the increase in F3 and F4 values, making them more prominent acoustically (see Figure 2.12).

Sound	Spectrogram
[a]	
[i]	
[u]	

Figure 2.12. *Spectrograms of the French vowels: [a], [i] and [u]*

When it comes to consonant sounds, several criteria come into play in order to describe their spectrograms. Criteria include the temporary transitions between formants, voicing[1], periods of silence, etc. Let us examine Figure 2.13, where we will consider some examples of consonants produced alongside the vowel [u].

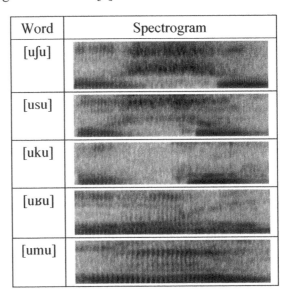

Word	Spectrogram
[uʃu]	
[usu]	
[uku]	
[uʁu]	
[umu]	

Figure 2.13. *Spectrograms of several non-sense words with consonants in the center*

As we can see in the spectrograms of Figure 2.13, the difference between the consonants [ʃ] and [s] can be seen by the position of the dark areas at the top of the spectrogram, reflecting the frequency of noise from friction which is much higher in the case of [s] than in the case of [ʃ]. The closure phase of the consonant [k] can be seen in the light area in the spectrogram followed by a little area of friction which corresponds to aspiration, a consequence of the release phase, since this is a voiceless consonant. The French fricative [ʁ] is seen as vertical bars of energy, which reflect friction. The nasal consonant [m] is characterized by a pattern that is similar to those of formants, shown as darker patches. Finally, the darker areas at the bottom of the spectrogram, which are visible for consonants [ʁ] and [m], reflect the fact that these consonants are voiced.

---

1 Voicing is the property of a sound which involves the harmonic vibration of the vocal folds.

2.1.1.3. *Auditory phonetics*

Auditory phonetics is the study of how sounds are perceived, looking at what happens within the ear and also what happens when the brain is processing sounds. The study of auditory phonetics fits within the frameworks of work in physiology and experimental psychology (see [JOH 11, RAP 11] for a general introduction).

Before looking at the perception of speech sounds, let us consider the anatomy involved in auditory perception. Instead of focusing on the complexities of the multimodal perception of speech sounds (which involve parallel auditory and visual pathways), we will focus solely on the auditory pathway; we recommend [MCG 76] and [ROS 05] for more information on multimodal perception.

The auditory pathway is made up of three main parts: the ears, the cochlear nerve and the brain. The *ear* is the main organ for hearing. It has three major functions: being stimulated by sounds, transmitting these stimuli and analyzing these stimuli. Having two ears enables the hearer to localize the source of sounds: a person speaking, the television, a car driving along, etc. The ear further away from the source receives the sound with a slightly weaker intensity and a slight delay compared to the other ear. Physiologically, three components make up the ear: the outer ear, the middle ear and the inner ear (see Figure 2.14).

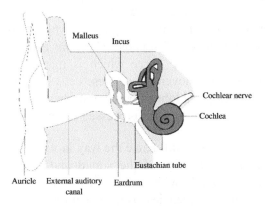

Figure 2.14. *Physiology of the ear*

The *outer ear* includes the auricle, whose role is to collect sounds and direct them towards the middle ear via the outer ear canal, which is about

2.5 cm in length. Sounds collected in this way then make the *eardrum* (a fibrous membrane located at the end of the ear canal) vibrate. Apart from its role as an amplifier, the eardrum plays an important role in the filtering out of certain frequencies and amplitudes.

The *middle ear*, which is located further in than the eardrum, is made up of a small hollow cavity which contains three ossicles that connect the ear drum with the cochlea. These ossicles include the *malleus*, the *incus* and the *stapes*. The role of these ossicles is the amplification of sound waves and the transmission of these sound waves to the inner ear. This mechanical transmission also ensures that that airwaves are transformed into liquid waves, transmitted via the fenestra ovalis. The difference in terms of the surface between the ear drum and the fenestra ovalis (a factor of 1/25) also contributes to the amplification of sound waves received.

Finally, the *inner ear* is made up of the cochlea. The cochlea is spiral shaped and has a diameter which progressively contracts. The cochlea is found inside a solid bony labyrinth which is full of fluid. The wave transmitted by the middle ear makes the basilar membrane inside the cochlea vibrate. This allows the initial analysis of sounds to take place, notably in terms of their frequencies, since the cochlea reacts to sound waves in a selective way. The lower part of the cochlea resonates with high-pitched sounds and the upper part (also called the apex) resonates with low-pitched sounds. The selectivity of the cochlea can be explained by the fact that diameter variation implies the existence of a continuum of natural frequencies which allows the cochlea to differentiate a great number of sounds. The *hair cells* within the cochlea are connected to the *cochlear nerve* which is made up of approximately 30,000 neuron axons. The movement of these cells triggers a signal which is directed towards the temporal lobe in both hemispheres of the brain. The brain merges these signals and the signal is processed and perceived.

Several perceptual factors influence the way in which we hear sounds, for example, the *pitch*, the quality of the sound and its length. An average person can perceive sounds with frequencies between 20 and 20,000 Hz. However, most sound frequencies used in natural languages are between 100 and 5,000 Hz. Since frequency variations are not always perceived in the same way by the human ear, we use the term *pitch*, measured on the *mel* scale, to describe the perceptual effect of frequencies.

The feeling of pitch, created by voiced sounds, such as vowels and consonants such as [b] and [d], is linked to the frequency of vocal fold vibration, which is also called the fundamental frequency F0. In the case of voiceless sounds such as [s] and [p], the sound produced is the result of the passage of air forced through constrictions. This produces faster pressure variations than with vowels. It is also worth mentioning that pitch variations play an important role in the marking of prosodic differences. When differences in pitch occur in a systematic way within words in certain languages (notably within a syllable) and create differences in the meaning of words as a result, we call these languages tone languages. Examples of such languages include Mandarin Chinese and Vietnamese. When pitch differences occur at the level of the whole phrase, we are talking about intonation, which is a phenomenon that occurs in many languages, including French, English and Arabic.

Volume (or *loudness*) is the perceptual equivalent of amplitude. Just as with the relationship between pitch and frequency, the relationship between volume and amplitude is not linear, because sounds of a very high frequency or a very low frequency must be of a much higher amplitude to be perceived (compared to sounds with a mid-range frequency). Typically, speaking more loudly is the normal reaction to problems where noise prevents perception. From a linguistic perspective, volume plays a minor role in the study of accents of other prosodic phenomena.

2.1.1.4. *The phonetic system of French*

Descriptive phonetics considers the global properties of linguistic systems (languages or dialects) as well as comparing them to find universals or phenomena that are specific to a given language. The description is carried out according to articulatory and acoustic criteria. These descriptions are particularly based on the distinction (universally attested in all of the world's languages) between consonants and vowels.

Vowels are voiced sounds (requiring the vibration of the vocal folds) which come from the passage of air in the oral cavity and/or nasal cavity without or with very little obstruction. To classify vowels, we use features such as manner of articulation, nasality, degree of aperture, as well as backness or frontness, which we will see in detail in the following sections.

The manner of articulation concerns the way in which the speech organs are configured to shape the air that comes from the lungs. This concerns phenomena such as rounding and orality/nasality. *Rounding* refers to the shape of the lips during the production of a vowel characterized by the pulling in of the corners of the mouth and the lips in the middle. It is accompanied by the protrusion of the lips, creating an additional cavity between the lips and the teeth which some call the labial cavity (see Figure 2.15).

Rounded   Unrounded

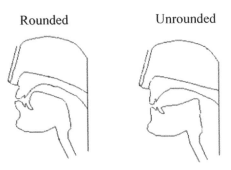

Figure 2.15. *Lip rounding*

Even though, in theory, all vowels can be pronounced with rounded lips, some can only be pronounced with rounded lips, such as the first group in Table 2.3 [ROA 02].

Rounded vowel	Example	Transcription	Unrounded vowel	Transcription	Example
[y]	vu<u>e</u>	[vy]	[i]	comm<u>is</u>	[kɔmi]
[œ]	P<u>eu</u>r	[pœR]	[e]	nomm<u>é</u>	[nɔme]
[ø]	P<u>eu</u>	[pø]	[ɛ]	<u>ê</u>tre	[ɛtR]

Table 2.3. *Examples of rounded and unrounded vowels in French*

Nasality is the result of the lowering of the velum, which forces the air to cross the nasal cavity. Since the mouth cannot be closed completely during the production of a vowel, air continues to pass (in parallel) through the oral cavity. In IPA, a tilde is added above the grapheme to mark nasal vowels. Several languages have nasal vowels, such as Portuguese, Polish, Hindi and French, which contains four vowels of this type (see Table 2.4).

Nasal vowel	Examples
[ɔ̃]	Mon, tronc
[ɑ̃]	Tente, gant
[œ̃]	Brun, défunt
[ɛ̃]	Satin, rien

Table 2.4. *Nasal vowels in French*

French also has a group of 12 oral vowels which are characterized by air only passing through the oral cavity, without nasalization (see Table 2.5).

Oral vowel	Examples
[i]	Ni, ici
[e]	Les, générer
[ɛ]	Mes, maison
[ø]	Peu, cheveux
[y]	Vu, mue
[ə]	Le, de
[u]	Cou, mou
[o]	Bateau, beau
[ɔ]	Fort, toc
[ɑ]	Pate, mât
[a]	Bas, ma
[œ]	Peur, sœur

Table 2.5. *Oral vowels in French*

The degree of aperture is another criterion which allows vowel sounds to be distinguished between one another. It is a question of the distance between the tongue and the palate at a place where the two are at their closest. To find out more about the differences between degrees of aperture, let us look at the following sequence of vowels:

1) [i] [y] [u]

2) [e] [ø] [o]

3) [ɛ] [œ] [ɔ]

4) [a] [ɑ]

As we can see, the first series corresponds to three close vowels because the tongue is close to the palate. The second series involves three mid-close vowels, and the third series includes three mid-open vowels. Finally, we have a series of two open vowels where the distance between the tongue and the palate is maximal.

Frontness/backness is an important criterion in the classification of vowels. Front vowels such as [i], [e], [ɛ] and [a] are characterized by the tip of the tongue being at the front of the mouth. On the other hand, back vowels such as [u], [o], [ɔ] and [ɑ] are characterized by the pulling back and bunching up of the body of the tongue (see Figure 2.16).

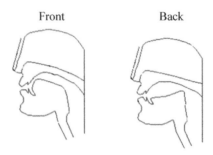

Figure 2.16. *Front vowels and back vowels*

Taking the criteria of openness and backness/frontness, we obtain the schematic classification of vowels which takes the form of a trapezium (see Figure 2.17).

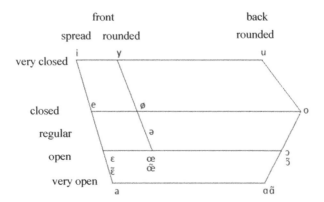

Figure 2.17. *French vowel trapezium*

It is useful to mention that there is a particular vowel in French which is called the silent *e*. It is transcribed using the [ə] symbol in IPA. This vowel is different to the vowel [ø], which is used in words such as deux [dø] (two). Indeed, both are front rounded vowels in French but with different degrees of rounding, [ø] being more rounded. Phonologically, this vowel has a special status in French because it can appear in precise contexts, notably at the end of a word or before another vowel. Furthermore, geographical considerations can come into play to regulate its pronunciation.

Consonants are sounds that are produced with a closure or a constriction in the oral cavity. In contrast to vowel sounds, the vibration of the vocal folds is not a necessary condition during the production of consonants. However, two main criteria are used to classify consonants: place of articulation and manner of articulation.

The *manner of articulation* of consonants allows two consonant sounds which share the same place of articulation but which are pronounced in different ways to be distinguished. This allows consonants to be classified according to the following pairs of categories: voiced/voiceless, oral/nasal, plosive/fricative and lateral/vibrant.

Voicing is the property of consonants whose pronunciation involves the vibration of the vocal folds. Thus, we have two series of consonant sounds: series (a) represents voiceless consonants and series (b) represents voiced consonant sounds.

a) [p] [t] [k] [f] [s] [ʃ]

b) [b] [d] [g] [v] [z] [ʒ]

In a practical way, to find out whether a consonant is voiced or voiceless, you can put your finger on your throat to feel whether there are vibrations.

As we have seen with vowel sounds, *nasalization* is a property which is the result of air passing through the nasal cavity due to the lowering of the velum (see Figure 2.18). In French, there are three nasal consonants: [m], [n] and [ɲ], like in *gagner* (*w*in). To test whether a consonant is nasal or not, you can block the nose while pronouncing this consonant. If the sound changes in quality, this shows that the consonant sound is nasal.

Oral   Nasal

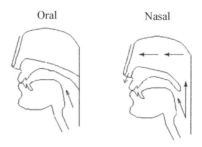

Figure 2.18. *Nasal and oral consonants*

Some consonant sounds are pronounced in a continuous and homogeneous way (through time), while others are quick, due to the complete closure of the vocal tract and then the sudden escape of air from this point of closure (like an explosion). This property allows *fricatives* (series a), which are continuous, to be distinguished from *plosives* (series b), which are quick.

a) [f] [v] [s] [z] [ʃ] [ʒ]

b) [p] [b] [t] [d] [k] [g]

From an articulatory point of view, the continuous aspect of fricatives is due to the constriction of the air passing through the vocal tract but without complete closure. Once the speech organs are in place for a fricative, they do not move. This gives fricative sounds a homogeneous character. However, the production of plosives requires the speech organs to move and, therefore, a heterogeneous sound is produced.

A lateral consonant is a consonant whose pronunciation involves air passing down the sides of the tongue. In modern French, there is only one lateral consonant: [l]. Conversely, a vibrating consonant is the result of one or more taps with the tongue tip or the back of the tongue making repeated contact against the uvula. For example, with the [r], which is commonly called apical, the tongue is placed against the upper teeth but produces a tap which allows air to pass through.

For consonant sounds, the place of articulation is the area where two articulators come close or touch one another, forming a partial or full closure of the vocal tract. This allows French consonants that we are taking as an example here to be classified according to Table 2.6:

Place of articulation	Phonetic category	Examples
Lips	Labial or bilabial	[p], [b], [m]
Teeth and lip	Labiodental	[f], [v]
Teeth	Dental	[t], [d], [n]
Alveolar ridge[2]	Alveolar	[s]
Palate	Palatal	[ɲ], [ʃ], [ʒ]
Velum	Velar	[k], [g]
Uvula	Uvular	[ʁ]

Table 2.6. *Places of articulation of French consonants*

There is a category of sounds that lies between consonant and vowel sounds and these are sometimes called *semi-vowels* or *glides*. These in French are non-syllabic vowel sounds (which cannot form the nucleus of a syllable) which, when next to syllabic vowels, form part of a diphthong, while also having linguistic properties which are similar to those of consonants. In some respects, semi-vowels could be defined as fricative consonants, which correspond to particular vowels sharing their place of articulation. The list of semi-vowels for French is given in Table 2.7.

Semi-vowel	Example	Corresponding vowel
[j]	Feuille [fœj]	[i]
[ɥ]	Puis [pɥi]	[y]
[w]	Bois [bwa]	[u]

Table 2.7. *French semi-vowels*

Finally, coarticulation is a phenomenon in phonetics which deserves to be mentioned. It occurs when a given sound influences the pronunciation of another sound or several sounds which follow or precede it. For example, in the case of a sequence of French words such as *robe serrée* (tight dress), the voiced *b* is influenced by the voiceless *s* and becomes a *p*. The sequence is pronounced [ʁɔpseʁe]. In this case, phoneticians talk about the assimilation of voicing.

---

2 The alveolar ridge is a cavity within the jawbone where the teeth are embedded [ENC 09].

2.1.2. *Phonology*

This is a branch of linguistics which looks at the functioning of sounds in the framework of a language system. In contrast to phonetics, phonology looks at sounds as abstract units and not as concrete physical entities. As a consequence, phonology adopts an abstract basic unit, the phoneme, not the physical sound adopted by phonetics. For a general introduction to phonology from a computational perspective, a good reference is [BIR 03].

2.1.2.1. *The concepts of phonemes and allophones*

The *phoneme* is a fundamental unit in phonology. It is the smallest unit in language which does not hold any meaning of its own, but changing one phoneme for another causes a semantic change. For example, the replacement of the phoneme /p/ in the word *pit* by the phoneme /b/ leads to a clear change in meaning, since the word *bit* is created. Since the words *pit* and *bit* only differ by one phoneme, we call them a *minimal pair*. These groups of words are particularly useful for identifying phonemes in a language. Sometimes, sound differences do not lead to semantic differences. This is the case notably with dialectal variations of the same phoneme, and the best-known case of this in French is with the phoneme /r/. It is well known that depending on the geographical zone, this phoneme is realized by the sounds [r], [R] and [ʁ]. Therefore, we are talking about three *allophones*. In other languages, these three sounds, or two of them as, in Arabic, correspond to three distinct phonemes. This demonstrates that two or more sounds can be allophones in one language and distinct phonemes in another.

It must also be outlined that although phonetics and phonology examine different aspects of speech, they share a number of things in common, for example the distinctive features that we will detail later (see [KIN 07] for more information about the connection between phonetics and phonology).

2.1.2.2. *Distinctive features*

Since the Middle Ages, linguists have recognized that sounds are not simple units but rather complex structures made up of phonetic characteristics. At the beginning of the 20th century, the Russian linguist Nicolaï Troubetzkoy described this aspect of phonemes in terms of oppositions and he identified several types of oppositions [TRO 69]. He described binary oppositions, which imply two phonemes that share properties in common, such as the phonemes /p/, and /f/, and exclusive oppositions, which show that in a pair of phonemes, only one might possess a particular trait (e.g. voicing in the case of /v/ and /f/, where only /v/ is voiced). He also described multilateral oppositions that concern several phonemes such as /p/, /b/, /f/ and /v/. Finally, he broached continuous oppositions which, by their very nature, vary in degrees such as the height of the tongue necessary for the production of a vowel.

The proposal for what is now known as the distinctive feature theory is historically attributed to Roman Jakobson whose work with Gunnar Fant and Morris Halle laid its foundations [JAK 61]. Jakobson's model stipulates, among other things, that all phonological features are binary. This means that a phoneme either possesses or does not possess a trait of whatever kind, therefore making the discretization of the gradual oppositions described by Troubetzkoy necessary. The advantage of this rule binarization is that it makes the expression of phonological rules much simpler. To guarantee the universality of its description, Jakobson opted for acoustic features, therefore avoiding articulatory features which are too dependent on specific language. The result of his taxonomy is a collection of 12 features: consonantal, compact/diffuse, grave/acute, tense/lax, voiced, nasal, continuant, strident (elevated noise intensity)/non-strident, obstruent, flat and sharp. For example, the phoneme /a/ possesses the following features: +Vowel, −Nasal, +Grave, +Compact.

Reexamined by Noam Chomsky and Morris Halle in their book *The Sound Pattern of English* (SPE), the concept of the distinctive feature becomes less abstract and gives way to a richer taxonomy. According to the SPE model, features are categorized according to five groups: major features, place of articulation features (cavity), manner of articulation features, source features and prosodic features [CHO 68] (see Table 2.8).

The description of phonemes in terms of features opens the path to generalizations in the form of phonological rules, which give way to the study of a number of morphological phenomena such as joining and assimilation.

Class	Features	Explanation
Major features	Syllabic [+Syll]	This feature indicates that the phoneme is capable of making up a whole syllable by itself. In French, for example, only vowels have this property, but in English, some consonants can be syllabic like the consonant *l* in *bottle*.
Major features	Consonantal [+Cons]	This feature concerns the phonemes whose production involves a major vocal tract obstruction such as plosives, liquids, fricatives and nasals.
Major features	Sonorant [+Son]	This feature indicates a significant opening of the vocal tract: vowels, semi-vowels, liquids and nasals.
Place of articulation	Coronal [+Cor]	This feature is connected to consonants where the tip of the tongue gets close to or touches the teeth or alveolar ridge.
Place of articulation	High [+High]	The body of the tongue rises to get close to or touch the palate.
Place of articulation	Anterior [+Ant]	The place of articulation is located at a frontal position in the mouth (alveolar ridge, teeth, lips, etc.).
Place of articulation	Back [+Back]	The place of articulation is located with the tongue behind the palatal area.
Manner of articulation of consonants	Sonorant [+Son]	This feature allows vowel and consonant phonemes which require the vocal and nasal tract to be as unconstructed as possible to be grouped together. Liquids and nasals are examples of consonants which have this feature.
Manner of articulation of consonants	Continuant [+Cont]	This feature allows sounds whose pronunciation can be prolonged (e.g. fricatives) to be distinguished from phonemes that cannot be prolonged (e.g. plosives).
Manner of articulation of consonants	Nasal [+Nas]	Phonemes whose pronunciation requires free airflow through the nasal cavity.

	High/Low [+High]/ [+Low]	This feature indicates the position of the tongue in the mouth, marking the degrees of aperture of the vowels: open [+Low], close [+High].
Manner of articulation of vowels	Back [+Back]	The body of the tongue is pulled back.
	Round [+Round]	Phonemes produced with rounded lips.

Table 2.8. *Examples of distinctive features according to the taxonomy by Chomsky and Halle [CHO 68]*

2.1.2.3. *Phonological rules*

This is a formal tool used most notably in the framework of generative phonology to describe a phonological or morphophonological process. The rules can apply to phonetic transcriptions or to feature structures. The modern format of these rules was proposed in the framework of the SPE model. Their general system has the following format: A → B / X_Y. In this system, X represents the position on the left and Y represents the right, while A and B correspond to the entry and exit symbols, respectively. In other words, after applying this rule, the sequence XAY will be replaced by the sequence XBY. To express higher level constraints, it is possible to eventually use the symbols $, +, # and ## to mark the boundary of a syllable, a morpheme, a word or a phrase, respectively. Likewise, the symbol ø, used to mark an empty element, is particularly useful for expressing the suppression of an element in the entry sequence.

To clarify the functioning of these rules, let us examine the following example:

$$[n] \rightarrow [m] / \underline{\quad} [b]$$

This rule describes the replacement of *n* by an *m* in standard Arabic if the final sound is followed by a *b*, independent of what is found in the left-hand position. For example, the *n* in the preposition من [min] (has several English equivalents depending on context including *of* and *from*) becomes *m* when it is followed by the adverb baʕd (after), which begins with a من بعد and the result is the following pronunciation [mimbaʕdi] ([ALF 89], see [WAT 02] for similar phenomena in Arabic dialects). It must be noted that a similar change occurs in Spanish but in different conditions, which require the description in terms of distinctive features:

$$[n] \rightarrow [m] \ / [+vowel] \ \underline{\qquad} \begin{bmatrix} +labial \\ +obstruent \end{bmatrix}$$

Following this rule, the phoneme *n* becomes *m*, if it is preceded by a vowel (such as [a], [e], [o], [i] and [u]) and followed by a consonant which is labial and an obstruent like [p], [b] and [f]. Thus, *n* does not change in a case like [en espaɲa] but it becomes *m* in cases such as:

– en Porto Rico [em portoriko], en Paris [em paris];

– en Bolivia [em boliβja], en Barcelona [em barθelona];

– en Francia [em fransja].

Phonological rules can occur in many different ways and they can be grouped into four categories: assimilation, dissimilation, insertion and suppression.

Assimilation involves changing the features of a given phoneme to make it phonetically more similar to another phoneme, which is usually adjacent but sometimes further away. The preceding two rules describe a case of consonantal assimilation under the effect of a neighboring consonant. The assimilation of the vowel [ɛ] in the French verb "aider" (to help) [ɛde] is a good example of vowel assimilation. This vowel is often pronounced [e] under the effect of the final vowel. In this case, since the two vowels are not adjacent, we can say that this is non-contiguous assimilation.

Dissimilation is a phonological phenomenon which has been shown both historically and in today's languages. In contrast to assimilation, it involves the systematic avoidance of the occurrence of two similar sounds being next to each other [ALD 07]. Let us look at the following examples from a Berber language [BOU 09]:

am + bur	[anbur]	Unmarried
am + frrd	[anfr:d]	Person who changes
am + mal	[anmal]	Action of showing

This gives us the following rule:

$$[m] \rightarrow [n] / \_\_\_ \begin{Bmatrix} b \\ f \\ m \end{Bmatrix}$$

Insertion involves adding a sound into a given phonological context. In English, a typical case involves adding the schwa [ə] at the end of words which finish with an *s* (before an s ending) such as in the word *bus* [bʌs], which becomes *buses* [bʌsəz] in the plural. Another example of insertion comes from Egyptians speaking English. The language spoken in Egypt does not allow three consonants to be pronounced in succession. Some Egyptian speakers of English will add a vowel sound to avoid three consonant sounds in succession. The sentence: *I don't know exactly* [aɪ doʊnt noʊ ɪgzæktli] becomes [aɪ doʊnti noʊ ɪgzæktili] with the insertion of the vowel *i* at the end of the words *don't* and *exact*.

Suppression can affect a given sound in a precise context where the speaker thinks that it will not cause an ambiguity if it is introduced. A typical case in French is the suppression (by certain speakers) of the schwa [ə] in words such as the verb "devenir" (to become) [dəvəniʁ] which might be pronounced [dvəniʁ] or the adjective "petit" (small), which is pronounced [pti]. Likewise, the last consonant in a word, like the final *t* in the adjective *petit* or the *s* in the determinant *les*, is not pronounced when it is followed by an obstruent, a liquid or a nasal. On the other hand, this consonant is pronounced when it is followed by a vowel or a semi-vowel. Let us look at the following examples:

Petit talon	[pəti talɔ̃]	Les talons	[le talɔ̃]
Petit lapin	[pəti lapɛ̃]	Les lapins	[le lapɛ̃]
Petit narval	[pəti naRval]	Les narvals	[le naRval]
Petit adolescent	[pətit adɔlɛsɑ̃]	Les adolescents	[lez adɔlɛsɑ̃]
Petit oiseau	[pətit wazo]	Les oiseaux	[lez wazo]

This gives the following rule:

$$[+\text{consonant}] \rightarrow \varnothing \,/ \,\underline{\quad}\, \# \left\{ \begin{array}{l} [+\text{obstruent}] \\ [+\text{liquid}] \\ [+\text{nasal}] \end{array} \right\}$$

In spite of the simplicity of these rules, using them as a representation framework is not unanimously agreed upon within the community, notably due to the ambiguity surrounding the order of their application. Different alternatives have been proposed to resolve this problem. One such alternative, which is the most easily applied in NLP, is two-level phonology. This theory will be the focus of a specific section in this chapter. Other approaches recommend using logic to maximize the rigor of the description framework, therefore minimizing the gap between the theory and its computational implantation [GRA 10, JAR 13].

2.1.2.4. *The syllable*

The concept of the syllable allows describing phenomena at a higher level than the phoneme level, which is the main focus of the SPE model. It involves an abstract phonological unit which corresponds to an uninterrupted phonetic sequence. As a phonological unit, a syllable can either have meaning or not.

The simplest approach involves characterizing the syllables in a language in a linear way, such as sequences of consonants (C) and vowels (V). Let us take the following as examples: a, V, the CV, cat CVC, happy CV-CV, basket CVC-CVC. As we can see, some syllables end with a vowel and are called open syllables (like CV), while others end with a consonant and are called closed syllables (like CVC).

Other theoretical approaches stipulate that the syllable should have a hierarchical structure [KAY 85, STE 88]. Following these approaches, a syllable (σ) is made up of two main parts: an onset (ω) and a rime (ρ). The onset is universal and present in all the world's languages. Although some languages allow empty onsets, they nevertheless require a structural marker to show this, often in the form of joining. For example, the syllable *y* [i] in French has an empty onset which is filled by the joining (liaison) phenomenon, like in "allons-y" (let's go (there)) [alõzi]. The rime is an obligatory part in every well-formed syllable. The rime is made up of two elements: the nucleus (v) and the coda (k). The nucleus is the most sonorous part of the syllable and is the underpinning element of the rime. This is why it is considered an obligatory element. In French, the nucleus is made up of a vowel, a diphthong, or both at the same time. However, some languages, including Czech, allow a nasal consonant or a liquid to fulfill this role. The coda is optional and has a descending sonority. It is made up of one or several consonants. Consequently, the syllables that have a coda are also closed syllables. Let us look at some examples in Figure 2.19 to illustrate some syllabic structures in French.

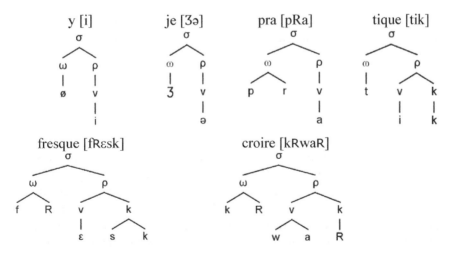

Figure 2.19. *Examples of some possible syllabic structures in French*

As we can see in Figure 2.19, the onset of the first syllable is the null element (ɸ). This is a way of marking that the onset is obligatory but that in this specific case, it is empty. Likewise, we can see that some constituents can form branches, like the onset in the syllable *pra*, the onset and the coda of the syllable *fRɛsk* or the nucleus of the syllable *croire*.

2.1.2.5. *Autosegmental phonology*

The SPE model has been criticized for several reasons. One reason is because it emphasizes the rules more than the substance of phonological representations. The other reason is the overgeneration of certain rules which are capable of producing unattested cases in the language whose system they are meant to be describing. This contributed to the emergence of different models including autosegmental phonology, an important extension of generative phonology, which was proposed by John Goldsmith also under the name *nonlinear generative phonology* [GOL 76, GOL 90] (see [PAR 93, VAN 82, HAY 09] for a general introduction).

This theory is based on abandoning the principle of the strict linearity of language introduced by Saussure in favor of a multidimensional vision of the linguistic sign previously defended by linguists from different branches of the field such as [BLO 48, FIR 48] and [HOC 55]. Thus, according to this theory, phonological representations are made up of several independent tiers, each having a linear structure.

The *segmental tier* focuses on distinctive features more than phonemes to give a finer granularity to the representation. For example, the phoneme /p/ is represented by the following features: [-sonorant, -continuant, -voiced and -labial]. To overcome the redundancy of the representations of these features, several models have been proposed to take the concept of underspecification between the features into consideration [MES 89, KIP 82, ARC 88, ARC 84].

The *timing tier* corresponds to the succession of units of time and is used as a pivotal point for the other tiers. It allows complex and long segments to be processed. Typically, these units are noted in the form of a sequence of characters, marked as *x* in the tree, and are associated with segments. For example, the consonant *m* which is doubled in the spelling of the French words "emmener" and "emmagasiner" receives two temporal units (see Figure 2.20).

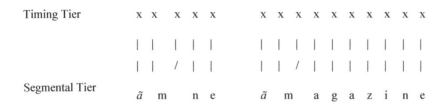

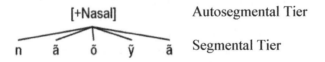

Figure 2.20. *Examples of how double consonants are dealt with by the timing tier*

The *tonal tier* contains information which shows the distribution of tones at the level of phonological representation. Several studies have shown that this tier is independent of segments since changes in elements at the segmental level do not affect the tonal structure that these elements carry. Tonal information is particularly important for tone languages such as Vietnamese and Chinese.

The *autosegmental tier* allows the study of long-distance vowel and nasal harmony [CLE 76]. This phenomenon is well known in certain South American languages, such as Warao which is spoken in Venezuela and Tucano and Barasana, spoken in Colombia. In these languages, nasality spreads from consonants to neighboring vowels like in nãõ +ya → nãõỹã, *he is coming* [PEN 00] (see Figure 2.21).

Figure 2.21. *Propagation of nasality in Warao*

The *syllabic tier* allows the constraints linked to the elements of the syllable that we saw in the section on the syllable (onset, nucleus and coda) in each language to be studied. The information in the syllabic tier corresponds to the root of the tree σ, while the segmental level corresponds to the phonemic sequence. As Paradis [PAR 93] highlights, differences in syllabic structure allow interesting phonological phenomena to be explored, such as joining (liaison) of words, e.g. *l'oiseau* [wazo] ("bird" in French), where the semi-vowel *w* is considered to be a part of the nucleus and not the onset, while this joining does not occur in *la ouate* [wat] ("wadding" in French), since *w* is part of the onset.

The *metrical tier* concerns the supra-syllabic level and allows phenomena such as the accentuation of one or more syllables, and other prosodic phenomena, to be described. This includes intonation that we will detail later (to avoid repetition of other prosodic phenomena).

As a theory of the dynamics of phonological representations, autosegmental phonology covers the rules about well-formedness, allowing the association of one tier's element with an element of another tier.

Autosegmental phonology has seen important developments, notably following McCarthy's work which promised to generalize the theory to take into account languages with non-concatenative morphology such as standard Arabic [McC 81].

2.1.2.6. *Optimality theory*

Proposed by Alain Prince and Paul Smolensky, optimality theory (OT) is a linguistic model situated in the framework of generative grammar. It states that the surface structure of languages is the result of the resolution of a certain number of conflicts between several constraints in competition [PRI 93, PRI 04]. The linguistic system is, therefore, considered as a system of conflicting forces. Although phonology is the field that OT fits into best, many works have attempted to apply its principles in other branches of linguistics, such as morphology, syntax and semantics (see [LEG 01, HEN 01, BLU 03]).

In contrast to the SPE model, the OT uses constraints more than rules. For example, in the SPE model, to express the fact that in Egyptian, three consonants cannot appear in succession, rules such as those provided below are used.

$$\left[+\text{con}\right]\left[+\text{con}\right] \rightarrow \left[+\text{con}\right]\left[+\text{con}\right]\left[\text{I}\right]/ \underline{\quad}\left[+\text{con}\right]$$

$$\left[+\text{con}\right]\left[+\text{con}\right] \rightarrow \left[+\text{con}\right]\left[+\text{con}\right]\left[\text{I}\right] / \underline{\quad}\#\left[+\text{con}\right]$$

In practice, this gives us cases like:

il bint kibiːra →il bint − **I** −kibiːra (the eldest daughter)

To process this case from the perspective of OT, we need to use the constraints that can be organized in a table (see Table 2.9).

Entrance	*CCC
il bint kibiːra	*
il bint − I − kibiːra 🖘	

Table 2.9. *Constraint forbidding three successive consonants in Egyptian Arabic*

As we can see in Table 2.9, the constraint of forbidding three successive consonants is violated by the first candidate (marked by an asterisk) and not by the second. This explains why the second candidate, marked by the pointing hand, is preferred over the other. If we take the different constraints implied in the formation of a sentence, we will find a number of candidates in competition and each will obey different constraints (but not all the constraints). The optimal candidate will, therefore, be the one that satisfies the most constraints.

To put candidates in order, OT distinguishes between different types of violations: a violation marked by an asterisk and a crucial violation marked by an exclamation mark. Let us examine the case of joining in French and constraints that are exemplified by two possible candidates [pəti ami] (boyfriend) and [pəti t ami] presented in Table 2.10. The constraints that apply to an example of this kind can be classified into two major categories: markedness constraints and faithfulness constraints. Markedness constraints imply a minimization of markedness and, therefore, cognitive effort, which contributes to a maximization of discrimination. In practice, this is realized by a preference for simple linguistic structures which are commonly used and easy to process. For example, all the languages in the world have oral vowels but few have nasal vowels. Likewise, all the languages in the world have words that begin with consonant sounds but few languages allow words to begin with vowel sounds. Identity constraints put structural changes upon entry at a disadvantage. In other words, they require words to have the same structure on entry as on exit: no insertion, no suppression, no order changes, etc.

As we can see in Table 2.10, we have two constraints which are in competition. Firstly, the constraint that two vowels cannot be next to each other is violated by the second candidate, but not by the first. We can also see that the faithfulness constraint is violated by the first candidate, which involves the insertion of a *t*. Since the first constraint is very important in French, its violation is considered to be fatal. Thus, the first candidate is favored, even though it violates the faithfulness constraint (see [TRA 00] for a more detailed discussion of similar examples in French).

Entry	*VV	Faithfulness
peti t ami 🖙		*
peti ami	*!	

Table 2.10. *Constraints involved in the case of joining (liaison) in French*

It can be said that the main advantage of OT is its explanatory ability. In contrast to the rules which must be satisfied by the surface structure, the constraints bring to light factors which might otherwise be hidden (not visible on the surface) but which contribute to the emergence of the optimal output structure.

From a computational point of view, the formal establishment and models of OT implicate other theoretical frameworks, such as autosegmental phonology used by Jason Eisner, who proposed a formalization of OT based on finite-state automata [EIS 97]. Based on Eisner's model, William Idsardi showed that the problem of OT generation is NP-hard and that it cannot give an account for phenomena of phonological opacity because it does not possess an intermediary level of rules to apply [IDS 06]. The Idsardi test was called into question by Jeffrey Heinz [HEI 09] who argued that it only applies to a particular part of OT. Boersma [BOE 01] proposed a stochastic view of OT where the constraints are associated with a value on a continuous scale rather than on a discrete scale of priority. Other formalization works have emerged in the field of syntax (see [HES 05] for a general review of the applications of OT in computational linguistics). Furthermore, a learning framework for OT was proposed by Bruce Tesar [TES 98, TES 12].

2.1.2.7. *Prosody and phono-syntax*

Prosody is a term in suprasegmental phonology used to discuss supraphonemic phenomena such as the syllable, intonation, rhythm and flow

[CRY 91, BOU 83] (see [VOG 09] for a typological overview). From a phonetic point of view, the parameters which allow prosody to be characterized include height, acoustic intensity and the duration of sounds. These parameters are directly affected by the emotional state of the speaker, which gives prosody an additional dimension.

In the field of NLP, prosody plays an important role in a number of applications, notably speech synthesis, emotion recognition, the processing of extra-grammatical information in oral speech, etc. Given the complexity of interactions between prosody and superior linguistic levels in French, we will concentrate on prosodic phenomena in this language, while also mentioning notable phenomena in other languages.

2.1.2.7.1. Stress

Stress has the function of establishing a contrast between different segments, defined by phonological, syntactic and semantic criteria, which are stressable units. As Garde [GAR 68] underlines, the criteria chosen to define these units vary from one language to another. Consequently, stress has the role of giving a formal marker to a grammatical unit, the word, which is an intermediary between the minimal grammatical unit, the morpheme and the maximal grammatical unit, which is the sentence.

Stress can be marked using two different types of processes: positive and negative. *Positive processes* involve the increase in intensity of the syllable, which is stressed in comparison with unstressed syllables. Another factor that contributes to a syllable being stressed is prolonging the length of the nucleus of the syllable in question, as well as a noticeable increase in its fundamental frequency. In contrast to positive processes, *negative processes* involve the removal of a feature in unstressed syllables. Logically, these processes affect the features that belong to the inventory of distinctive features in the language.

In French, *stress units* have limits which are highly variable and depend on factors such as the succession of syllables which are susceptible to being stressed, the rhythm of speech and pauses. The majority of words are susceptible, in certain positions, to becoming unstressed. In reality, stress does not affect units that can be described based on fixed linguistic criteria. It is more a question of units whose limits vary from one utterance to another. This is how French tends to avoid the immediate juxtaposition of two stresses. Apart from this case, French excludes the possibility of semantically close words all being stressed in near succession unless there are pauses. Groups of words like this become longer and the flow becomes gradually more rapid and less careful. Given that there is no syntactic category that systematically carries the stress in French, it is impossible to syntactically define a stress unit. On the other hand, we can define the syntactic constraints which affect the realization of stress. For example, we know that the stress cannot be placed on an article such as *le* or *la* (the), since these morphemes are always integrated into a larger stress unit.

In some languages, the position of the stress, in the framework of a stress unit which has already been set out, depends upon syntactic criteria. These languages are called *variable stress* languages. In other languages, called *fixed stress* languages, the position of the stress depends on phonological criteria. In French, stress is fixed, because it falls on the final syllable in the stress unit.

Now, let us consider an important variant of ordinary stress in French which is called *insistence stress*. Just as with ordinary stress, it is based on the idea of an intentional contrast which allows certain elements of the utterance to become the focus of the utterance. In contrast to ordinary stress, this insistence stress falls on the first syllable of the word in French and this causes perceptibility to increase. In French, it is possible to distinguish two processes in prosodic insistence: emotional insistence stress and intellectual insistence stress [GAR 68]. *Emotional insistence stress* involves prolonging the first consonant of a word with an emotional value or pronounced with disapproval, like in *C'est formidable* (it's wonderful) or *C'est épouvantable* (it's terrible). *Intellectual insistence stress* involves reinforcing intensity by increasing the fundamental frequency as well as lengthening the first syllable of the stressed phrase [GAR 68]. It is mainly used to mark the opposition between two terms like in *c'est un chirurgien qui a opéré le patient, pas un infirmier* (it was a surgeon who operated on the patient, not a nurse) or in the following exchange:

– *Un billet d'avion avec une chambre d'hôtel ?* (A plane ticket with a hotel room?)

– *Non, un billet d'avion seulement.* (No, just a plane ticket)

The position of the intellectual insistence stress is not always the same as in the case of emotional insistence stress. Both are on the initial syllable when the word begins with a consonant. On the other hand, in words which begin with a vowel sound, intellectual insistence falls on the first syllable and emotional insistence falls on the second.

2.1.2.7.2. Intonation

Intonation, in the broadest sense of the term, covers a whole series of physical parameters which vary with time, such as intensity, fundamental frequency and silence. Psychoacoustic parameters also come into play with intonation. Stress, melody, rhythm, prominence, breaks, etc. are all important [ROS 81, ROS 00, GUS 07].

Since the beginning of linguistics, there have been a number of studies investigating the relationships between syntax and intonation, given the possible range of effects of syntax on the surface structure of the phrase. Intonation plays a role in the initial resolution of syntactic ambiguities and, therefore, allows the speaker to choose a particular analysis over other analyses or interpretations. Let us examine the following sentence: John talked #1 about his adventure #2 with Tracy. Depending on the place of the pause (1 or 2), one possible interpretation is preferred. The possible interpretations are as follows, respectively: Tracy went on the adventure with John, or Tracy is John's conversation partner. Here is another interesting example: John didn't leave his house # because he was ill. With a pause after the prepositional phrase, the interpretation of this sentence is: John is at home. Without the pause, the interpretation becomes: John is no longer at home but the reason for his having left the house is not the illness. Finally, it is well known that intonation plays an important role in the marking of discursive structures and thematic roles [HIR 84] (see [HIR 98] for a typological review).

2.2. Speech processing

Speech processing (SP) is a term which refers to a variety of applications. Some applications are limited to the level of the speech signal, whereas others rely on high-level linguistic information. In this section, we will focus on the two applications which fit into NLP the best: speech recognition and speech synthesis. Given the nature and the objectives of this book, we will not elaborate on the aspects linked to the processing of the signal itself, but, nevertheless, we will include references for those who would like to know more.

Before going any further, we must quickly present two other applications which are of interest in many respects: speaker recognition and language identification.

Speaker recognition involves identifying the person who utters a phonic sequence whose length varies from long to very short. Such applications require vocal characteristics (which are unique to each and every one of us) to be modeled. To limit the quantity of speech required for training, a generic model is created and the specific models for each speaker are derived from this. Once the models have been created, the process of identification involves measuring the distance between the short sequence produced by the speaker and the existing models. The speaker is identified if the distance between the sample and the model exceeds the pre-established confidence threshold (see [BON 03, KIN 10] for an introduction).

Another application of SP involves identifying the language or dialect of the speaker. From a practical point of view, such applications allow telephone calls to be connected to speakers who are able to understand and interact with the language used, in an international context. However, applications to the field of security remain the most common for this type of systems. The principle of these systems is similar to the systems of speaker identification. This involves creating an acoustic profile for each target language and/or dialect and measuring the distance between the sample received and each of the existing profiles (see [MON 09] for an example of these works).

2.2.1. *Automatic speech recognition*

Automatic speech recognition (ASR) involves identifying sequences of words which correspond to the speech signal captured by a microphone (see [JUN 96, YU 14] for a general review). The most obvious use for ASR is in the context of human–machine spoken dialogue applications, where data are introduced by means of spoken utterances. Nevertheless, vocal dictation is the oldest and most widespread application of ASR.

To classify ASR systems, several parameters come into play, including the size of the vocabulary and the number of speakers. Table 2.11 presents the main parameters as well as the main characteristics of the systems [ZUE 97].

Parameter	Possibilities
Speaking mode	Isolated words, continuous speech
Speaking style	Read text, spontaneous speech
Enrollment	Dependent or independent speaker
Vocabulary size	Between 20 and tens of thousands of words
Type of language model	Finite state machine or context-dependent
Perplexity	Between < 10 (small) and >100 (large)

Table 2.11. *Classification parameters of speech recognition systems*

Given the complexity of their task, ASR systems have to overcome big challenges. These challenges have given way to highly desirable characteristics, which are:

– robustness: an ASR system must adapt to different levels of sound quality, including sounds of a poor quality. This poorness of quality can be the result of noise in the environment. This is especially the case in applications that are designed to be used in a car or in an airplane (e.g. the noise caused by the engine or the air). Conversations happening nearby can also present a similar form of challenge;

– portability: given the cost of the development of an ASR application, it is highly desirable to be able to apply the same work to several areas of the application without much effort being required;

– adaptability: a good system should be able to adapt to speaker changes and microphone changes and this should be the case whenever it is used. Spontaneous speech comes with its own major challenges to overcome, as it contains hesitations and repetitions which are difficult to model, even with the use of statistical approaches. Finally, a recognition system must be capable of recognizing or at least reducing the impact of the words pronounced which are not already included in its vocabulary.

Typically, a speech recognition system is made up of several modules and each one is specialized, allowing it to deal with a particular aspect of the analysis of the speech signal (see Figure 2.22).

Firstly, the *feature extraction module* allows features which are useful for the digital and sampled signal to be extracted. In telephone applications, for example, this sampling is carried out at a rate of approximately 8,000 samples per second. In some cases, this module has the role of improving the quality of the signal, by reducing the sources of sound which come from a neighboring conversation, from noise in the environment, etc. This filtering allows the recognition module to be activated intelligently when it is established that the signal received is effectively a speech signal and not noise. For more information on the extraction of features and signal processing in general, refer to [CAL 89, OGU 14] and [THA 14].

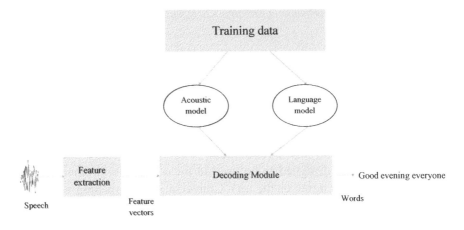

Figure 2.22. *General architecture of speech recognition systems*

Next, the decoding module tries to find the most likely word using acoustic models and language models.

The acoustic model allows hypotheses about words to be generated with the help of techniques such as hidden Markov models (HMM), neural networks, Gaussian models or other models. The result of the acoustic model is a graph of words which sometimes involves a great number of sequences, and each one corresponds to a path in the graph. To identify the best sequence of words in the graph, the decoding module uses a language model which is usually based on n-grams. Note that in spite of the domination of statistical models since the 1970s, different forms of pairing with linguistic models have nevertheless been experimented with.

2.2.1.1. *Acoustic models based on HMM*

When work first began on ASR, rule-based expert systems were used to detect phonemes. Given the limits of their efficiency and the difficulties surrounding their development, these approaches were abandoned in favor of statistical methods, notably Markov chains. Markov chains are mathematical systems proposed by the Russian mathematician Andrei Markov at the beginning of the last century to model temporal series. The model was then further developed by the American mathematician Leonard E. Baum and his collaborators whose work gave way to the HMMs. The first implementations of HMM in the area of speech recognition took place during the 1970s by Baker at Carnegie Mellon University, as well as Jelinek and his colleagues at IBM [BAK 75a, BAK 75b, JEL 76].

The fundamental idea of Markov chains is to examine a sequence of random variables which are independent of each other. The purpose of such a sequence is that it allows us, through observing past variables, to predict the value of these variables in the future. For example, if we know the monthly rainfall in a given town, we can predict, with a certain margin of error, what the rainfall will be in the subsequent month(s). Therefore, we are talking about conditional probability, where the value of a given variable (quantity of rain predicted for the subsequent month) depends upon the values of variables in the sequence that precedes it (known history of rain in the town in question).

For an introduction to the fundamental concepts of probability, since this goes beyond the objectives of this text, we recommend [KRE 97] and [MAN 99] (for an introduction to these concepts in the context of NLP). A great range of books and manuals giving an introduction to probability exists and, of course, these can be used if necessary.

Let us take a detailed example to explain the principle of the Markovian process (see [RAB 89, FOS 98] for a similar example on climate). Suppose that a given person or robot, whom we will call Xavier, can be in three possible moods: happy, neutral and sad. Suppose also that our Xavier remains in the same mood for the whole day. Therefore, to predict Xavier's mood tomorrow, it is necessary to know what his mood is today, what it was like yesterday and what it was like the day before yesterday, etc. More formally, we must calculate the probability of $p(e_{n+1}|e_n, e_{n-1}, \ldots e_1)$. In practice, it is best to take into account a part of Xavier's mood history with only a limited number of preceding moods. Here, we are talking about a first-order Markov chain (with only one preceding mood) or a second-order Markov chain (with two moods preceding the current mood), etc.

Suppose that we have a chain of random variables $X = \{X_1, X_2, .., X_T\}$ and each one takes a value v based on a limited range of values $V=\{v_1, v_2, .., v_n\}$. This gives us the following equations:

$p(e_{t+1} = v_k | e_t, e_{t-1}, \ldots e_1) = p(e_{t+1}|e_t)$ first-order Markov process.

$p(e_{t+1} = v_k | e_t, e_{t-1}, \ldots e_1) = p(e_{t+1}|e_t, e_{t-1})$ second-order Markov process.

The probability of a Markov chain can be calculated according to equation [2.1].

$$p(e_1, \ldots, e_n) = \prod_{i=1}^{n} p(e_i \mid e_i - 1) \qquad [2.1]$$

Let us imagine that in the case of Xavier's moods, we have the probabilities presented in Table 2.12.

Mood today	Mood tomorrow		
	Happy	Neutral	Sad
Happy	0.70	0.27	0.03
Neutral	0.40	0.35	0.25
Sad	0.12	0.23	0.65

Table 2.12. *Probabilities of Xavier's moods tomorrow, with the knowledge of his mood today*

Table 2.12 shows that the probability of a radical change in Xavier's mood (e.g. happy→ sad) is generally inferior to that of a gradual change (e.g. happy→ neutral) or to no change at all (e.g. happy→ happy). The information given in this table can be represented in the form of a Markovian graph made up of a finite number of states (which is three in our case – namely neutral, happy and sad), of transitions between these states (arrows or curves) which allow a transition between one state (mood) and another, as well as the probability of staying in the same state (Figure 2.23).

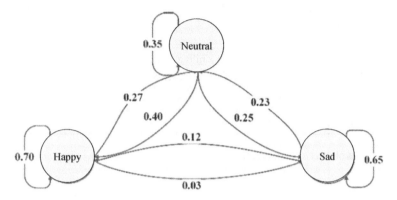

Figure 2.23. *Markovian model of Xavier's moods*

This model allows us, for example, to calculate the probability that Xavier's mood will be neutral tomorrow and happy the day after tomorrow, knowing that his mood is neutral today. This is carried out in the following way:

$p(e_2=\text{neutral}, e_3=\text{happy} \mid e_1=\text{neutral}) =$

$p(e_3=\text{happy} \mid e_2=\text{neutral}, e_1=\text{neutral}) * p(e_2=\text{neutral} \mid e_1=\text{neutral}) =$

$p(e_3=\text{happy} \mid e_2=\text{neutral}) * p(e_2=\text{neutral} \mid e_1=\text{neutral}) = (\text{first order Markov})$

$0.4 * 0.35 = 0.14$

The idea of Markov chains was pushed even further following the work by Baum and his collaborators who added an additional element of complexity to the model, which is latent or hidden variables, and this gave way to what we today call the hidden Markov models or HMMs. The main purpose of HMMs compared to ordinary Markov chains is that they allow sequences to be dealt with even if they contain ambiguity and, as a consequence, they can be processed in more than one way.

If we come back to our example, we know that in reality it is very difficult to guess if someone is in a good mood or not (hidden variable). The only way of telling this is to observe the person's behavior (observable variable). We can simplify this into two possible behaviors: smiling or frowning. This allows us to imagine the probability presented in Table 2.13 containing information about whether Xavier is likely to be smiling or frowning, knowing his mood.

Mood	Probability of smiling	Probability of frowning
Happy	0.88	0.12
Sad	0.55	0.45
Neutral	0.65	0.35

Table 2.13. *Probability of Xavier's behavior, knowing his mood*

A more detailed version of the Markov chain graph (Figure 2.23) with behavioral probability gives the HMM diagram shown in Figure 2.24.

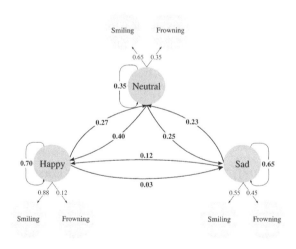

Figure 2.24. *HMM diagram of Xavier's behavior and his moods*

As Figure 2.24 shows, the HMM model is based on two types of probabilities: the probability of transition and the probability of emission. Transition relates to the movement from one state to another while emission involves emitting an observable variable based on a certain state.

From a formal point of view, HMMs can be seen as a quintuplet:

$E = \{e_1, e_2, ..., e_n\}$	All states
$O = \{o_1, o_2, ..., o_m\}$	Observations
$\pi(i) = p(e_i=ti)\ 1 \leq i \leq n$ where n is equal to the number of states	Probability of being in a state e_i at time t knowing that $t = 0$ for the initial state
$A = \{a_{ij}\}$ with $i, j \in E$	Probability of transition between states
$B = \{bijo\}\ i, j\ \{E\ and\ o\}$	Probability of emission
$\lambda = \{A, B, \pi\}$	Complete model

If we come back to our example, the HMM allows us to calculate the probability of Xavier's mood, which can be observed directly, based on his behavior, which is also observable directly. In other words, it is possible to calculate a particular mood $e_i \in \{happy, sad, neutral\}$ based on an observed behavior $o_i \in \{smiling, frowning\}$. This is carried out with the help of the Bayes formula, given in equation [2.2]:

$$p(e_i|o_i) = \frac{p(o_i|e_i).p(e_i)}{p(o_i)} \qquad [2.2]$$

In practice, we are sometimes more interested in the probability of a sequence of events than the probability of a single event. Thus, for a sequence of n days, we will have a sequence of moods $E=\{e_1, e_2, ..., e_n\}$ and a corresponding sequence of observable behaviors $O=\{o_1, o_2, ..., o_n\}$:

$$p(e_1,..e_n|o_1,...,o_n) = \frac{p(o_1,...,o_n|e_1,..e_n) \cdot p(e_1,..e_n)}{p(o_1... o_n)} \qquad [2.3]$$

When applied to the problem of ASR, the sequence of acoustic parameters extracted from the speech signal can be described by a HMM.

This involves combining two statistical processes: a Markov chain which models the temporal variations, and an observed sequence which allows spectral variations to be examined. The most intuitive way of representing a phonemic sequence is to consider that each state corresponds to a phoneme. Let us examine the example of the French word *ouvre* (to open) [uvʀ], which is made up of three phonemes, shown in Figure 2.25.

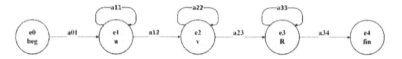

Figure 2.25. *Markov chain for the word "ouvre" (open)*

As we can see in Figure 2.25, there are two types of transitions between phonemes: a transition between a phoneme and the following phoneme and a cyclical transition towards the same phoneme which allows important variations between the length realizations of each phoneme to be studied. These variations are generally due to the nature of the phoneme (notably continuous vowels and consonants), the differences in the context of phonetic realization, intraspeaker variation or interspeaker variation, etc. Of course, this model is highly simplified because the system in this case can recognize only one word. In applications that are slightly more complex, we have a limited number of words to recognize, like with vocal command systems, where we can give a limited number of orders to a robot or some kind of automatic system. For example, the Markov chain used for the recognition of three (French) vocal commands, namely *ouvre* (open), *ferme* (close) and *démarre* (begin) can have the form shown in Figure 2.26.

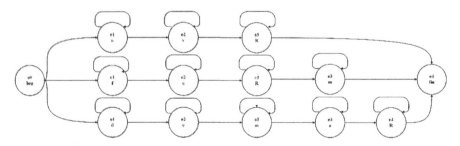

Figure 2.26. *Markov chain for the recognition of vocal commands*

In continuous speech recognition systems with a large vocabulary, a more narrow representation of phonemes is indispensable. Thus, we use several states to represent phonemes. This allows acoustic variations of the same phoneme within a spectrogram to be studied (see Figure 2.27).

Neural networks are a possible alternative for the classification of phonemes (see [HAT 99] for a review). In spite of their results, which are comparable to those of HMM, this technique suffered from significant limitations such as the fact that the learning process is quite slow and that it is difficult to estimate its parameters. Following recent developments in the field of deep learning, which found the solution to these limitations [HIN 06, HIN 12], we are seeing neural networks grow stronger as a real alternative to HMMs.

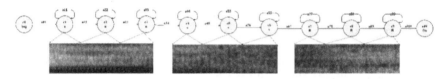

Figure 2.27. *HMM for the word "ouvre" (open)*

2.2.1.2. *The Viterbi algorithm*

As we have said, the role of HMMs in an ASR system is to help find the most likely sequence of phonemes, knowing the acoustic parameters extracted from the speech signal. This comes back to finding the most likely sequence of states or phonemes $E = \{e_1, e_2, \ldots, e_n\}$, knowing the acoustic parameters observed $O = \{o_1, o_2, \ldots, o_m\}$ and a model λ. In other words, what we are looking for is: $p(O, E|\lambda)$, which can be calculated using Bayes' equation (equation [2.4]).

$$p(O, E|\lambda) = p(O|E, \lambda) \cdot p(E|\lambda) \qquad [2.4]$$

Likewise, the probability of an observed sequence O knowing the HMM model λ can be calculated by the following equation:

$$p(O|\lambda) = \sum_E p(O, E|\lambda) \qquad [2.5]$$

To find the most likely path, knowing an observed sequence in a trellis, the easiest solution, but the least efficient, involves calculating the probabilities of every path that allows the sequence in question to be generated, and to then choose the most likely path. A brute force approach like this collides with the size of the research space where the number of possible sequences exponentially increases in the following way: $|E|^n$, where $|E|$ is the number of hidden states in our HMM model and n the size of the entry sequence. In real applications, where the number of states can exceed 100 and the number of words to be recognized can easily exceed 10, we can imagine that the number of possibilities to consider can quickly become impossible to deal with in this way.

The most popular approach for calculating the most probable path is to use the Viterbi algorithm. This algorithm, proposed at the end of the 1960s by Andrew Viterbi, the Italian American engineer [VIT 67], is based on dynamic programming. This paradigm involves dividing the original problem into sub-problems whose solution leads to the solution of the whole problem. The algorithm uses two variables:

– $\delta_t(i)$ is the path for which the likelihood is maximal among all possible paths and ends with the state s_i at time t:

$$\delta_t(i) = \max_{e1,\ e2,\ ...,\ et-1} p(e_1,\ e_2, ...,\ e_{t-1},\ e_t = s_i,\ o_1,\ o_2, ...\ o_t \mid \lambda)$$

– $\psi_t(i)$ allows to store the best path, ending in state s_i at time t:

$$\psi_t(i) = \operatorname*{argmax}_{e1,\ e2,\ ...,\ et-1} p(e_1,\ e_2, ...,\ e_{t-1},\ e_t = s_i,\ o_1,\ o_2, ...\ o_t \mid \lambda)$$

The principle of the Viterbi algorithm involves finding the most probable path for each intermediary state and later using it to find the final state in the trellis. Thus, at each time t, we come back to the most probable curve which leads to s_i. The algorithm is made up of four steps:

1) Initialization:

$\delta_1(i) = \pi_i \cdot b_{i,o1}$, $i = 1, .., T_s$ where T_s is the number of states and π_i the probability of being between a state e_i at moment $t=1$.

$\psi_1(i) = 0$ No state precedes the initial state.

2) Recursion:

The recursive function is the true heart of the algorithm. Informally, the problem of the final path is cut up into sub-problems whose partial solution is stored in variable ψ_t for the current state j:

$$\delta_t(j) = \max_{1 \leq i \leq Ts}(\delta_{t-1}(i) \cdot a_{ij}) \cdot b_{j, e_n} \qquad\qquad 2 \leq t \leq T$$
$$1 \leq j \leq T_s$$

$$\psi_t(j) = \arg\max_{1 \leq i \leq Ts}(\delta_{t-1}(i) \cdot a_{ij}) \qquad\qquad 2 \leq t \leq T$$
$$1 \leq j \leq T_s$$

3) Stop condition:

The algorithm stops once we arrive at the final state T:

$$P^*(O|\lambda) = \max_{1 \leq i \leq Ts} \delta_T(i)$$

$$e_T^* = \arg\max_{1 \leq i \leq Ts} \delta_T(i)$$

Next, we look for the best path when the end of the observation sequence has been reached at moment $t = T$, starting with the best vectors ψ_t.

4) Path backtracking:

$$E_t^* = \{e_1^*, .., e_T^*\} \text{ such as: } e_t^* = \psi_{t+1}(e_{t+1}^*), t = T - 1, T - 2, \ldots, 1.$$

To illustrate the functioning of this algorithm, let us go back to the example of Xavier's behavior. Suppose that we want to calculate the most probable sequence of states, knowing that we have observed the following sequence of behaviors: O = {smiling, smiling, frowning}. Since we do not know Xavier's initial mood: t_0, we will assume that the three possibilities are equally probable. To simplify the presentation, we will adopt the following conventions: Happy = ha, Sad = sa, Neutral = nt, Smiling = sm and Frowning = fr.

1) Initialization:

$t = 1$

$$\delta_1(ha) = \pi_{(ha)} \cdot b_{ha,sm} = 0.33 * 0.88 = 0.29$$

$\psi_1(\text{ha}) = 0$

$\delta_1(\text{nt}) = \pi_{(\text{nt})} \cdot b_{\text{nt,sm}} = 0.33 * 0.65 = 0.21$

$\psi_1(\text{nt}) = 0$

$\delta_1(\text{sa}) = \pi_{(\text{sa})} \cdot b_{\text{tr,sm}} = 0.33 * 0.55 = 0.18$

$\psi_1(\text{sa}) = 0$

2) Recursion:

We then calculate the likelihood of being in a certain state based on three possible predecessors, from which we choose the most likely.

$t = 2$

$$\delta_2(\text{ha}) = \max(\delta_1(\text{ha}) \cdot a_{\text{ha.ha}},\ \delta_1(\text{nt}) \cdot a_{\text{nt.ha}},\ \delta_1(\text{sa}) \cdot a_{\text{sa.ha}}) \cdot b_{\text{ha, sm}} =$$
$$\max\left(0.29*0.70, 0.21*0.40, 0.18*0.12\right) \cdot 0.88 = 0.178$$

$\psi_2(\text{ha}) = \text{ha}$

$$\delta_2(\text{nt}) = \max(\delta_1(\text{ha}) \cdot a_{\text{ha.nt}},\ \delta_1(\text{nt}) \cdot a_{\text{nt.nt}},\ \delta_1(\text{sa}) \cdot a_{\text{sa.nt}}) \cdot b_{\text{nt, sm}} =$$
$$\max\left(0.29*0.27, 0.21*0.35, 0.18*0.23\right) \cdot 0.65 = 0.050$$

$\psi_2(\text{nt}) = \text{ha}$

$$\delta_2(\text{sa}) = \max(\delta_1(\text{ha}) \cdot a_{\text{ha.sa}},\ \delta_1(\text{nt}) \cdot a_{\text{nt.sa}},\ \delta_1(\text{sa}) \cdot a_{\text{sa.sa}}) \cdot b_{\text{sa, sm}} =$$
$$\max\left(0.29*0.03, 0.21*0.25, 0.18*0.65\right) \cdot 0.55 = 0.064$$

$\psi_2(\text{sa}) = \text{sa}$

$t = 3$

$$\delta_3(\text{ha}) = \max(\delta_2(\text{ha}) \cdot a_{\text{ha.ha}},\ \delta_2(\text{nt}) \cdot a_{\text{nt.ha}},\ \delta_2(\text{sa}) \cdot a_{\text{sa.ha}}) \cdot b_{\text{ha, fr}} =$$
$$\max\left(0.178*0.70, 0.050*0.40, 0.17*0.12\right) \cdot 0.12 = 0.014$$

$\psi_3(\text{ha}) = \text{ha}$

$$\delta_3(nt) = \max(\delta_2(ha) \cdot a_{ha.nt}, \, \delta_2(nt) \cdot a_{nt.nt}, \, \delta_2(sa) \cdot a_{sa.nt}) \cdot b_{nt, \, fr} =$$

$$\max\left(0.178*0.27, 0.05*0.35, 0.17*0.23\right) \cdot 0.35 = 0.016$$

$$\psi_3(nt) = ha$$

$$\delta_3(sa) = \max(\delta_2(ha) \cdot a_{he.sa}, \, \delta_2(nt) \cdot a_{nt.sa}, \, \delta_2(sa) \cdot a_{sa.sa}) \cdot b_{sa, \, fr} =$$

$$\max\left(0.178*0.03, 0.05*0.25, 0.17*0.65\right) \cdot 0.45 = 0.005625$$

$$\psi_3(sa) = nt$$

At this stage, as Figure 2.28 shows, we have a trellis that is fully populated with three possible paths which contain values of variable ψ as follows:

$$\psi_3(ha) = ha, \, \psi_2(ha) = ha,$$

$$\psi_3(nt) = ha, \, \psi_2(nt) = ha,$$

$$\psi_3(sa) = nt, \, \psi_2(sa) = sa,$$

Thus, the first path is: {happy, happy, happy}, the second is: {happy, happy, neutral} and the final one is: {sad, neutral, sad}.

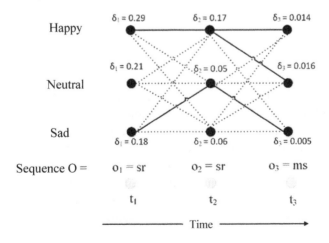

Figure 2.28. *Trellis with three possible paths*

3) Stop condition:

We begin to look for the most probable sequence based on the previous stage:

$$P^*(O|\lambda) = \max(\delta_3(e)) = \delta_3(nt) = 0.016$$

$$e_3^* = \operatorname{argmax} \delta_3(e) = nt$$

This means that the most probable state with t=3 is neutral.

4) Backtracking:

Based on the previous state found in the preceding stage, we return backwards to the previous state: $t = T-1 = 3-1 = 2$.

$$e_2^* = \psi_3(e_3^*) = \psi_3(nt) = ha.$$

Again, we come back to a state: $t = T-1 = 2-1 = 1$.

$$e_1^* = \psi_2(e_2^*) = \psi_2(ha) = ha.$$

This means that, knowing the sequence of behaviors observed: O= {smiling, smiling, frowning}, the most likely sequence is E= {happy, happy, neutral}. Note that with the probabilities of transitioning between states only, the most likely path would be {happy, happy, happy}.

2.2.1.3. *Language models based on n-grams*

After the phoneme recognition phase, ASR systems must find words that correspond to the sequence of phonemes obtained in the preceding phase. Of course, a sequence of letters can correspond to a multitude of possible word sequences. This is due to several factors, such as homophony (two different words with the same pronunciation) or phoneme recognition errors, etc. For example, the (French) phonetic sequence [vjɔlɔ̃sɛlki] can correspond to "*violoncelle qui*" (cello that) as well as "*violons celle qui*" (violate that one). In the same way, in English, based on the phonetic sequence [naɪ.treɪt], it is possible to obtain two different word sequences: "*night rate*" and "*nitrate*". Furthermore, it must be indicated that language models are used in several applications outside of ASR such as machine translation, information retrieval, text classification, etc.

To find the optimal sequence of words, knowing the sequence of phonemes, several sources of information come into play, notably morphology and syntax. Traditionally, syntactic grammars were used to refine the search for a word sequence. That was abandoned in favor of more statistical approaches sometimes combined with certain forms of syntax [JEL 99]. However, the ideal combination of discrete syntactic knowledge with statistical knowledge, or other forms of continuous knowledge, is yet to be found as much at the level of linguistic theory as at the application level.

From a formal point of view, the probability of observing a sequence of words $p(w_1, ..., w_m)$ is approximated according to the following equation:

$$p(w_1, ..., w_m) = \prod_{i=1}^{m} p(w_i \mid w_1^{m-1}) \approx \prod_{i=1}^{m} p(w_i \mid w_{i-(n-1)} ..., w_{i-1}) \qquad [2.6]$$

The idea of n-grams is to consider the *n-1* preceding words rather than the complete history. The hypothesis is that partial information loss is justified by the gain from the reduction in the quantity of linguistic data necessary for n-gram learning. The size of the data required for learning increases with the value of *n*. The general case of n-grams can be calculated using the following equation:

$$p(w_n \mid w_1 ... w_{n-1}) = \frac{p(w_1 ... w_n)}{p(w_1 ... w_{n-1})} \qquad [2.7]$$

To fully understand n-grams, we are going to use the following micro-corpus where sentences begin and end with the markers <p> and </p>, respectively:

<p>I am Mr. Dujardin</p>
<p>It's Dujardin</p>
<p>I want a single room for one night with a balcony </p>

Several types of n-grams exist following the size of the history that is taken into consideration. *Unigrams* are the simplest form of n-grams. It is a matter of counting the frequency of words out of context in a given corpus, without taking past frequencies into account. However, the raw information from the frequency of a given word in a corpus is difficult to interpret, hence the use of normalization in the form of probability. Thus, we proceed to the division of the frequency of the word in question by the total number of words in the corpus (equation [2.8]):

$$p(w_x) = \frac{f(w_x)}{C} \tag{2.8}$$

where C is the number of words in the corpus and $f(w_x)$ is the frequency of the word w_x. If we begin to count the words in our micro-corpus, we obtain Table 2.14.

Word	$f(w_x)$	$P(w_x)$	Word	$f(w_x)$	$P(w_x)$
I	2	0.11	a	2	0.11
am	1	0.05	room	1	0.05
Mr	1	0.05	single	1	0.05
Dujardin	2	0.11	for	1	0.05
it	1	0.05	night	1	0.05
is	1	0.05	with	1	0.05
want	1	0.05	balcony	1	0.05

Table 2.14. *Micro-corpus unigrams*

Bigrams involve considering a given word with a back window of a single word. Thus, we are researching different possibilities of occurrences of pairs of words, that is: $f(w_x \; w_y)$. Let us take the example of the bigrams of our micro-corpus presented in Table 2.15. It is necessary to add n-1 artificial symbols (hashtags for example) at the beginning and at the end of each sentence to indicate the symbols which are located at the head and tail of the sentence. In the bigram example, the markers <p> and </p> played this role.

Bigram	f(w$_x$ w$_y$)	Bigram	f(w$_x$ w$_y$)
\<p\> I	2	single room	1
I am	1	room for	1
am Mr	1	for a	1
Mr Dujardin	1	a night	1
Dujardin \</p\>	2	night with	1
\<p\> It's[3]	1	with a	1
It's Dujardin	1	a balcony	1
want a	1	balcony \</p\>	1
a room	1		

Table 2.15. *Bigrams in the micro-corpus with their frequencies*

To normalize the bigrams, we can use the following equation:

$$p(w_n| w_{n-1}) = \frac{f(w_n \; w_{n-1})}{f(w_{n-1})} = \frac{p(w_n, \; w_{n-1})}{p(w_{n-1})} \qquad [2.9]$$

Based on Table 2.15, we can calculate the probabilities of certain bigrams as follows:

$P(I|\text{<p>}) = 2/3 = 0.6$ $P(am|I) = 1/2 = 0.5$ $P(balcony|a) = 1/2 = 0.5$

To calculate the probability of the sequence *It's Dujardin*, we can proceed as follows: $p(\text{<p>} \text{ It's Dujardin } \text{</p>}) = p(\text{It's}|\text{<p>}) \; p(\text{Dujardin}|\text{It's}) \; p(\text{</p>}|\text{Dujardin}) = 1/3 * 1/1 * 2/3 = 0.217$.

Finally, since it is not realistic to assume that the training corpus includes all possible words from the language in question, it is highly recommended to create in advance a specific category for unknown words.

---

3 For simplification, we have considered "*it's*" as one word.

2.2.2. *Speech synthesis*

A speech synthesis (SS) system is a software or hardware capable of producing speech in an artificial way [TAY 09]. In some cases, the input for such systems can be a phonetic transcription of the entry text or a text annotated with syntactic or prosodic information. However, since the typical input into these systems is a flat text, they are sometimes called *Text To Speech* systems.

Science historians have reported several attempts to create mechanical machines equipped with spoken functionalities such as Wolfgang von Kempelen's machine, which was created towards the end of the 18th century. However, the first SS system in the modern sense of the term was probably the VOCODER system, which was developed towards the end of the 1930s at the Bell laboratory in the United States.

From an application point of view, SS systems can be used as independent applications or as modules in larger systems. Applications for the visually impaired or the blind as well as for people with speech disorders are probably the most obvious medical use of such systems. Likewise, human–machine dialog systems use SS modules as a way of interacting with humans. Furthermore, with the increased popularization of intelligent tutoring systems, notably in the field of foreign language teaching, the use of an SS module gives a voice to virtual tutors, allowing them to teach the language in its spoken form. Another important development is the emergence of talking heads or animated virtual characters. There must be a detailed synchronization between the SS and the module that creates the animation of facial and body expressions to guarantee a minimal level of coherence. Of course, the success of all applications depends on the quality and expressiveness of the SS module.

A typical example of the architecture of an SS system is shown in Figure 2.29. Of course, it is more of a general example since the real implementations of modules as well as their modes of interaction can change.

As Figure 2.29 shows, in general an SS system involves two major steps: *front-end* processing, whose function is to prepare the entry text for synthesis, and *back-end* processing, which relates to the synthesis of the digital signal. This is done according to several approaches, like concatenative synthesis, formant synthesis and articulatory synthesis.

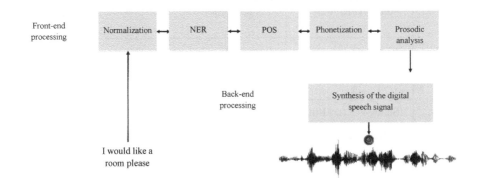

Figure 2.29. *Typical architecture of an SS system*

2.2.2.1. *Front-end processing*

This phase is made up of several stages of low-level linguistic processing, and the majority of these are explained in the following chapters. The first step is the normalization phase, which involves translating written conventions and abbreviations in an explicit format. Let us examine the examples from Table 2.16. As we can see, normalization is far from being a trivial operation. A simple syntactic conversion from one format to another is not enough, but it is clear that we need to access the meaning of the expression to translate it correctly.

The annotation of words by POS tagging (e.g. verb, noun, adjective, etc.), which we will see in detail in Chapter 3, allows certain pronunciation problems linked to homographs of different morphological categories. A typical example in English would be *live* whose pronunciation as a verb is clearly different from its pronunciation as a noun. In French, the three words *couvent, violent* and *relations* for example, can all be nouns or verbs with a different pronunciation for each category. For instance, in the case of *couvent* (convent (N) or brood (V)), the POS analysis allows the verb (pronounced [kuv]) to be differentiated from the noun, which is pronounced [kuvɑ̃]. However, this annotation does not resolve cases where homographs appear in the same grammatical category. The noun *bass* is a typical case in English as it may have two different pronunciations ([bas] and [beys]) depending on its meaning. As an example in French we have the word *fils* (son) [fis] and *fils* (threads) [fil].

Abbreviated form	Longer form
€1.25	one euro twenty five
3/5	three fifths
1/2 L milk	half a liter of milk
Sandy is 5 ft. 11 in Sandy is 5' 11"	Sandy is five feet and eleven inches tall
1 h 30, 1.30 h	one hour and a half or one hour thirty
Washington Ave. Delaware Cir.	Washington Avenue Delaware Circle

Table 2.16. *Abbreviations to be normalized before synthesis*

Named-entity recognition allows chains of words used to designate a specific object to be detected. It can act upon the title of a book or a film, the name of an institution, a company, a town, etc. Although according to the normative grammar of language such objects should be placed between quotation marks, in practice, many people do not follow this rule, and this makes it necessary to detect entities by non-orthographic means. Let us examine the following examples:

I saw Gone with the Wind last night.   Title of a film.

I would like to see Les Intouchables again with   Film title in
my school friends.   French.

As we can see from the above examples, the pronunciation of a named entity must be differentiated from the rest of the sentence to guarantee that it will be considered as such and not as a part of the sentence, in which case it would change the meaning. In cases where the name of the entity is in another language, a special form of treatment is preferred. For example, the proper noun (of Arabic origin) Hamza is pronounced [amza] in English, while in Arabic, it is pronounced [ħamza].

To find the sequence of phonemes based on words, we have the choice between two approaches: a symbolic rule-based approach (written manually or created automatically based on a corpus) and a dictionary-based approach.

The rule-based approach has the advantage of being able to process general cases without a need to store in the computer memory all possible forms. For example, the conversion rules of the grapheme *c* into *s* or *k* can be expressed as follows:

c → s / ___ {e, i, y}   *c* is transcribed as *s* if it is followed by one of the following French vowels: e, i, y.

c → k   *c* is transcribed as *k* in all other cases.

Another typical example in French is the pronunciation of the letter *g* which has two equivalents: [g] and [j] (see [BOS 97, MAN 00] for examples of learning-based approaches).

In the dictionary-based approach, all known possibilities are stored in a database. The advantage of this approach is that it allows the numerous pronunciation exceptions (notably in languages where pronunciation is not very regular, such as English) to be taken into consideration. The CMU pronouncing dictionary, which is freely available on the Internet, is a good example of one of these dictionaries[4]. It is an electronic dictionary for North American English with 125,000 words transcribed phonetically (stress marks included). The transcription format adopts the Arpabet code and includes 39 phonemes. The vowels are annotated with a number which indicates the stress that they carry. Thus, 0 indicates that there is no stress, 1 indicates a primary stress and 2 indicates a weak secondary stress which tends to be used in compound words such as *Vacuum Cleaner*, for example. Some transcription examples are given in Table 2.17.

Phoneme	Word	Transcription without stress marks	Transcription with stress marks
AA	odd	AA D	AA1 D
AE	at	AE T	AE1 T
AH	hut	HH AH T	HH AH1 T
IH	it	IH T	IH1 T

Table 2.17. *Examples of transcriptions with the Arpabet format*

---

4 http://www.speech.cs.cmu.edu/cgi-bin/cmudict.

The problem is that a dictionary, whatever its size, is limited and cannot cover all the words in a language, especially neologisms. To limit non-covert cases in the dictionary, some researchers proceed to the concatenation of compound words or morphemes. For example, imagine that we have to process the word *bavardation*, which is not listed in a French dictionary. We could assume that we are looking at a case of concatenating the French root *bavard* (chatty) and the suffix *ation*. Likewise, the English neologism *boatness* can be the result of the concatenation of the word *boat* with the suffix *ness*.

Finally, based on the above information, the *prosodic analysis module* allows the text to be analyzed to detect the units which are capable of receiving prosodic forms. The expression of emotions is an important subject, which is directly linked to prosody. The same utterance pronounced with different stress patterns leads to different emotional charges being attached, and this can sometimes change the meaning of the utterance (see [BUR 15] for a review of this work).

2.2.2.2. Concatenative synthesis

Concatenative synthesis is the most intuitive method because it involves putting recorded sequences of speech end to end. Besides its simplicity, it has the advantage of producing a good quality of speech, even though the passage from one sequence to another sometimes causes audible noise. The question which is posed here is one of optimal granularity. Of course, it is tempting to record relatively long sequences of speech such as entire sentences. Such units are well adapted for applications which do not require a large amount of variation in the message, such as indicating the next subway stop. In this case, in a simplified context, it is possible to imagine that it is sufficient to record a station announcement sequence (and the sequence of station names) like in the following example:

<station_announcement><little_pause><station_name_X>.

< station_announcement > = The next station is

< station_name _01>= Lexington Ave.

< station_name _02>= Broadway

etc.

Since the majority of applications require much larger variations in terms of messages, an extreme case being the synthesis of open texts, we must think about using more fine-grained units. The other extreme, in terms of granularity, involves recording the phonemes and concatenating them. Although this approach seems to respond to the need for message variability, it does not consider contextual phonetic variations caused by coarticulation, and therefore, considerably reduces the quality and the intelligibility of the speech.

To take coarticulation into consideration at a local level, larger units are used such as diphones, triphones or nphones, for general cases. Some researchers, like Thierry Dutoit, call this method *segmental unit selection speech synthesis* [DUT 00]. A diphone is an acoustic segment which begins at the center of the stable zone of a phoneme and ends at the center of the stable zone of the next phoneme. In the case of non-continuous phonemes, it is possible to take the most stable part or simply use triphones [DUT 97]. If the number of phonemes in a given language is equal to N, the number of diphones in this language is theoretically equal to N^2. Since not all theoretical combinations are authorized by linguistic systems, the real number of diphones is much smaller. For example, the 25 phonemes of Japanese give way to 625 diphones and the 40 phonemes of English produce 1,600 diphones but in both languages, only one part of the possible diphones is practically usable [BLA 00].

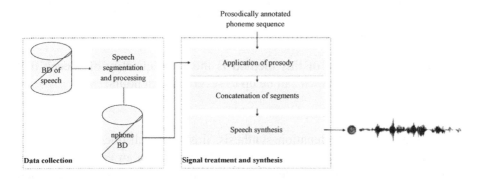

Figure 2.30. *General architecture of a concatenation synthesis system*

As Figure 2.30 shows, the first step in the process of constructing a SS system consists of collecting linguistic data in the form of a speech database, which must be representative of all the diphones of the target language. The simple approach involves retaining only the example of diphones which is judged to be the most representative. These sequences include all possible combinations of consonants, vowels and silence, which can be considered as a phoneme in the model. A more elaborate approach involves storing several examples of diphones with different prosodic properties, notably in terms of pitch and duration. This allows prosodic processing to be reduced to a minimum The cost of this reduction to increase of the size of the database.

Next, the signal retained for the base is edited and segmented with the help of specialized software, the most common one being Praat, developed by Dutch phoneticians Paul Boersma and David Weenink[5]. To facilitate adding prosodic parameters and to reduce the size of the database, it is necessary to carry out signal sampling with a method like linear predictive coding. Among the most popular systems developed with this technique, we can cite the MBrolla system from the Mons Polytechnic in Belgium as well as the Festival system developed at the University of Edinburgh in the UK and Carnegie Mellon University in the United States [DUT 96, LEN 00].

To guarantee the best quality of synthesis, industrial approaches adopt more varied natural units, which are generally superior to diphones, such as phonemes, syllables or morphemes. On the one hand, it allows the coverage of coarticulation phenomena on a broader scale. On the other hand, it also reduces the acoustic and prosodic treatments, thus guaranteeing better quality speech production. To find these stored units in an efficient way, an index is created with information such as acoustic characteristics, duration or the phonemic context of each unit.

The main downside of this method is the large quantity of speech that needs to be recorded, which can be up to several dozen hours.

2.2.2.3. Formant synthesis

In contrast to concatenation synthesis, this approach does not require a data sample to be stored because speech is produced in an entirely artificial way, based on an acoustic specification of the first formants of the sound [KLA 80]. Each formant is typically described using three acoustic parameters, namely frequency, amplitude and bandwidth, which represent the breadth of the spectrum signal surrounding the formant. In the majority of approaches, the oral and nasal cavities are modeled in a parallel way and then combined with a propagation module which simulates the nose and the

---

5 http://www.fon.hum.uva.nl/praat/.

lips. This particular method is not interested in modeling the speech production process by humans, but rather in modeling its entry and exit. In other words, the human phonatory system is considered a black box.

In a general way, this approach uses the source–filter model where the source models the activity of the vocal folds and the filter represents the vocal tract, which produces the formants. The sound source for vowels and consonants can take different forms according to their phonetic characteristics. Thus, the source for vowels and voiced consonants is either a periodic function or the output of a linear time-invariant filter, which takes a sequence of impulses as its input. Sounds implying an obstruction are generated based on random noise (white noise), while voiced fricatives use two sources. Formants are then created based on a second-order filter, which reduces the high frequencies and the amplitude from their source. Two types of architectures are possible: serial architecture and parallel architecture (see Figure 2.31). Serial architecture requires the frequency and the bandwidth of each resonance, as well as the common gain. This architecture, which is capable of approximating oral sounds, presents limits when it comes to approximating nasal sounds or fricative sounds. Parallel architecture is capable of approximating any kind of speech spectrum but requires the individual gains at the input of each filter.

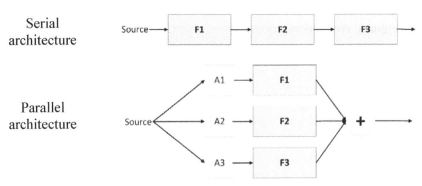

Figure 2.31. *Serial and parallel architecture of formant speech synthesis systems*

This technique offers multiple notable advantages. Apart from the good intelligibility of the speech that it produces, it allows the memory necessary for the system to be reduced, as it does not store speech. Furthermore, since speech is produced in a completely artificial way, it gives a very high level of control to parameters such as prosody and flow. The only disadvantage is probably that speech synthesized using this method can be quite robotic in some cases.

2.2.2.4. *Articulatory synthesis*

Articulatory synthesis is a technique which is based on the computational modeling of the physiological process of speech production. According to this approach, the sound produced is the result of the interaction of the vocal source with a filter that corresponds to one or several speech organs (see [PAL 06] for a general review). In other words, rather than specifying the desired speech signal, the conditions which lead to its formation are described. In this kind of model, artificial articulators are typically controlled by a collection of articulatory parameters whose variation with time allows a speech signal to be obtained. These parameters can be the geometry and the movement of the relevant articulatory organs (such as the tongue, the lips or the palate) as well as the forces and temporal information linked to these organs. To create these models, different methods can be used, including static methods (e.g. X-rays) or dynamic methods (e.g. electropalatography, electromagnetic articulography or optopalatography).

This approach has the advantage of being easily configurable and is, therefore, very promising. Today, its focus is on improving our comprehension of the language production process.

Morphology Sphere

3.1. Elements of morphology

Morphology is a branch of linguistics that focuses on the way in which words are formed from morphemes. Morphemes are the smallest linguistic unit which hold any meaning; thus, a word can be composed of one or more morphemes. For example, *book*, *cat* and *house* are simple words, while *books*, *finished* and *dismiss* are complex words because they are made up of multiple morphemes. There are two types of morphemes: lexical morphemes and grammatical morphemes. Also called *monemes*, lexical morphemes designate common objects such as *book, computer, city* or *flight*. These morphemes are distinguished by their number in a given language; it is always possible to add to the list of lexical morphemes with new morphemes, which are generally referred to as neologisms. Grammatical morphemes, on the other hand, concern words that play a grammatical role in a sentence, such as prepositions, articles and pronouns. Because these groups of words cannot in practice be modified by speakers of a language, they constitute closed groups. Thus, it is possible to add a new noun to designate a new object, but not to add a new preposition or pronoun.

The consideration of morphology as an independent branch of linguistics like phonology, syntax and semantics is not unanimous within the linguistic community. Certain syntactic theories, such as distributed morphology, posit that the role of syntax is to make all of the combinations required to construct a sentence, in terms of both word construction and the formation of syntactic units [HAL 93]. In addition, there are many interconnections between phonology and morphology. As we saw in the previous chapter, several phonological phenomena manifest during the realization of a phonological marker such as the plural. The relationship between semantics and morphology does not need to be proven since the meaning of an utterance is a function of the meanings of the morphemes that compose it.

In this chapter, we will try to cover key morphological properties of several langauges, such as English, French, Arabic and Turkish. For a general overview of morphology and its theoretical tendencies, we refer readers to [VIL 93, LIB 09] and [KIR 04].

3.1.1. *Morphological typology*

In terms of morphological typology, there are two major language groups: isolating languages and inflected languages.

Isolating languages, also called analytic languages, are languages (such as Vietnamese and Chinese) in which the morpheme-to-word ratio is very close to 1:1. In other words, in the majority of cases, the words in these languages are formed of a single morpheme, which means that their form hardly changes at all. These languages are rare and replace morphological information with the syntactic context represented by the order of words/morphemes. For example, a word such as *prehistoric* is equivalent, in these languages, to the following word/morpheme sequence: pre histor ic.

Inflected (or synthetic) *languages* constitute a linguistic family in which there is a principal morpheme, the stem, surrounded by affixes. This group can be further divided into two subgroups: fusional languages and agglutinative languages.

In *fusional languages* such as Arabic, words are not modified by adding morphemes before or after the stem, but rather by inserting new phonemes within the stem itself. Take the example of the Arabic stem k-t-b ("to write") which gives a multitude of derived words (see Table 3.1).

Arabic word	Translation	Arabic word	Translation
Kataba	wrote	Kateb	writer
Kaataba	correspond	Maktaba	library
Kitab	book	Maktab	office
Kutub	books	Maktoub	written

Table 3.1. *Examples of Arabic words derived from the stem k-t-b*

Agglutinative languages are another type of inflected languages. Unlike fusional languages, these languages form words from a number of morphemes that are completely separate from one another. For example, in the Turkish language, which is the most representative of this group, words begin with a stem followed by one or more suffixes [OFL 94]. The difference between this and other inflected languages is that it is possible to construct entire sentences with this procedure because the suffixes can be prepositions, possessives, etc. (see Table 3.2 for examples).

Word	Morphemes	Translation
sütsüz	süt-süz / milk-without	without milk
bankadan	bank-adan / bank-from	from the bank
gitti	gitt-i / went-he	he went
akarsuyunuz	akar-su-yunuz / river-2PP-POSS	our river is

Table 3.2. *Examples of words in Turkish*

Finally, we should mention that the classification presented above does not apply to all languages as there are languages, such as Japanese, which show mixed characteristics.

3.1.2. Morphology of English

Throughout the different phases of its development, English has been influenced by linguistic sources that have left their mark on its morphological system. Besides Latin and Greek, which form the classical substratum, English words have been borrowed from languages as varied as French, Spanish, German and Arabic. This has contributed to the creation of a rich lexical and morphological system.

3.1.2.1. *Types of morphemes in English*

English words are generally composed of a stem and an optional set of affixes. The stem, as a morpheme that cannot be removed, is the true morphological base of an English word. For example, in a plural word such as *books* we can remove the morpheme *–s* but not the morpheme *book*.

Stems may be surrounded by multiple secondary morphemes called affixes. These are classified into one of three types according to their placement in relation to the stem: prefixes, suffixes and circumfixes.

Prefixes are morphemes that precede the stem. In the vast majority of cases, they do not exist independently of the stem. Their role is essentially semantic, in that they do not affect the grammatical role of the word. Over time, some prefixes may become part of the lexicon, and consequently, be used independently as stems, like the prefix *ex*, which means ex-husband or ex-wife when used alone rather than as part of a word. In some cases, multiple prefixes may be placed before the stem, as in *anticonstitutional* in which the prefixes *anti* and *con* are used. Other examples are provided in Table 3.3.

Prefix	Meaning	Examples
im–	opposite	impossible, imprudent, imparity
pre–	before	prehistoric, premeditate, premature
anti–	Against	anti-abortion, anti-war, antibiotic
extra–	beyond	extrasensorial, extracellular, extrarural
over–	too much	overcook, overdose, overkill
para–	self	automobile, autobiography, autocracy

Table 3.3. *Examples of prefixes commonly used in English*

Suffixes are morphemes that come after the stem. Unlike prefixes, their role is not limited to modifying the meaning of the word but extends to the modification of its grammatical function. For example, if we add the suffix *–ly* to the adjective *slow*, we get the adverb *slowly*. For this reason, we classify suffixes according to the grammatical category they give; thus, there are nominal suffixes such as *–tion* and *–ment* and adjectival suffixes like *–ful* and *–al*. In Table 3.4 are provided some examples of common suffixes.

suffix	Meaning	Examples
– ist	Person who practices an activity	specialist, guitarist, linguist
– er, –or	Agent of an action	employer, actor, lawyer
– ism	Doctrine, belief, practice	sophism, abolitionism, animism
– able	Able to be	reasonable, portable, affordable
– ac	Pertaining to	cardiac, paranoiac, insomniac
– al	The action or process of	remedial, denial, trial

Table 3.4. *Examples of suffixes commonly used in English*

Finally, in bound or amalgamated morphemes, the affix is inseparable from the stem. For example, we cannot separate the stem from the suffix in verbs such as *began* or *bought*. Because these are conjugated verbs, we know that this is a case of a stem followed by an ending that is inseparable from it.

3.1.2.2. Allomorphs

Similar to the concept of an allophone, an allomorph involves phonetic or graphic variants of the same morpheme. In other words, this is a case in which two chains of phonemes or characters (depending on whether we are speaking of the oral or written case, respectively) correspond to the same morpheme. These variations may be contextual or combinatorial. In this case, the context in question may be phonological in nature, the plural morpheme –*s* is probably the most obvious case. In fact, in a word like *hats* it is realized without modification [s], while it is realized as [z] in words like *dogs* and finally as [əz] in words such as *boxes*.

3.1.2.3. Morpheme combination operations

There are several morphological combination operations we can use to form words from phonemes in English; these include inflection, derivation, composition and blending.

Inflection (or inflexion) is an operation that consists of combining the stem with a grammatical morpheme to give number for a noun, or time and person for a verb. They belong to the same category as the original, and the semantic difference between the base word and the new word, obtained via inflection, is easily noticeable. The purpose of inflection is, thus, essentially syntactic, such as gender or number agreement. For example, the following pairs of words are created via inflection: *livre/livres* (plural), *veuf/veuve*, (feminine), *mangent/mangeons* (conjugation).

Like inflection, derivation consists of combining the stem with a grammatical morpheme. It is distinguished by the fact that it produces a more significant semantic change. In addition, words composed by derivation generally belong to a different syntactic category than the starting word. For example, via the derivation process, we obtain the adverb *constitutionally* from the adjective *constitutional*, which is itself constructed from the noun *constitution*, which is itself obtained from the verb *constitute*. In other words, in this case, we have the following chain of words: verb → noun → adjective → adverb.

Composition, or compounding, is another type of word creation, in which at least two stems are juxtaposed within the same word. From an orthographical point of view, these words are written using three conventional methods which have no linguistic motivation. The first method consists of stringing together morphemes without spaces or hyphens, as in *darkroom*, *railroad* and *smalltalk*. In the second method, words are composed of stems separated by a space, as in *ice cream* and *real estate*. Finally, in the third method, words are composed of stems separated by a hyphen, as in *merry-go-round* and *actor-director*. There are several criteria which allow us to identify compound words. For example, nothing can be inserted between the parts of a compound word. Likewise, the semantics of compound words is a special case, since the overall meaning of a compound word is not equal to the assembled meanings of its components.

Blending, is a procedure that consists of joining the start of one word to the end of another. For example, the word *mockumentary* is obtained in the following way: mock + documentary. These words are usually used in the fields of science, technology, advertising and even poetry, in order to create nouns specific to new objects or concepts. Common examples include internet (international network), malware (malicious software) and bionic (biology + electronic).

3.1.2.4. *Abbreviation*

Abbreviation consists of creating a word by simplifying one or more other words. Truncation is a simplification procedure that consists of removing one or more syllables, as in <u>labo</u>ratory, <u>math</u>ematics, <u>Pat</u>rick,

Edward, etc. Another form of simplification consists of using the initials of the words in an expression or a complex noun. Words obtained in this way are called acronyms. Some of these acronyms are widely known, such as EU (European Union) and NASA (National Aeronautics and Space Administration), while others, like EMNLP (Empirical Methods on Natural Language Processing) and LDC (Linguistic Data Consortium) are known only in specialized fields.

3.1.3. *Parts of speech*

In the context of syntactic parsing, it is also useful to identify the part of speech of each word in the text or sentence under consideration. Given the many ambiguities that can exist, this procedure is anything but trivial. In reality, it is often necessary to use a combination of morphological structure and syntactic context in order to attain this objective, sometimes extending to include the semantic and practical levels. This explains why some specialists consider part-of-speech annotation to be part of syntax rather than morphology.

The classic categorization of many European languages is an inheritance from Latin and goes back to the beginning of the last century. It posits the existence of 9 categories, a view that is currently being questioned in general linguistics (see [CRE 95]). We will avoid these controversies and give a general description of the various categories and their basic syntactic properties.

The parts of speech adhere to the general classification of the language's words into two categories: open categories and closed categories. Open categories, the members of which cannot be limited, include the categories: noun, verb, adjective, and adverb. The closed categories, on the other hand, include so-called functional categories such as determiners, prepositions, conjunctions, interjections, and numerals.

Along with verbs, nouns are universally acknowledged as categories that must be present in all known languages. Nouns in English vary in number but unlike languages such as French and Arabic they don't have gender (feminine or masculine). Besides, there are nouns in English that do not vary in number, such as *sheep* and *salmon*. Some invariant nouns can sometimes have a plural form as well. For example, *antelop* and *antelopes* can be used as plural of *antelop* the same goes for fish that can have *fish* or *fishes* as plurals. Some nouns are always singular, like *physics*, *news* and *furniture*, while others, like *trousers* and *scissors*, are always plural. Based on syntactic and semantic criteria, we can distinguish two types of nouns: common nouns, also called substantive nouns; and proper nouns. Substantive nouns are almost always preceded by a determiner, and sometimes by an adjective, which can also be placed after the noun. In terms of syntax, the noun, or more precisely the noun phrase of which it is the core, plays a variety of roles, including subject, object, attribute or complement of another noun (see example [3.1]).

The <u>teacher</u> gave a gift to the best student.	Subject
He likes his <u>children</u> a lot.	Object
He is the <u>boyfriend</u> of Melissa.	Attribute [3.1]
His passion for reading is so intense that he would spend the entire day in the library.	Complement of another noun

In French, proper nouns have also a gender, and in certain exceptional cases may vary in number. However, this causes a semantic change, as in the group of proper nouns in example [3.2].

a) Paris est très jolie.	Paris is very beautiful (+ mark of fem.)
b) Seattle est très verte.	Seattle is very green (+ mark of fem.)
c) L'Amérique est en mouvement	America is moving
d) Les Amériques	The Americas [3.2]
e) Les États-Unis	The United States (federation)
f) Les Indes	Indias (old name for India)
g) L'Inde	India (modern India)

In terms of syntax, the particular characteristic specific to proper nouns is that they can form a noun phrase alone or plays its role at least. Although usually used alone, a proper noun can also be preceded by a determiner, as in [3.3].

– The Netherlands	
– The Rolling Stones	[3.3]

In English, adjectives are invariable while they vary in both gender and number in French depending on the nature of the noun to which they are connected. In English, there are three major types of adjectives: attribute, predcative and nominal. Attributive adjectives are used before (prepositive) or after (postpositive) a noun, typically playing the role of head of the nominal phrase. Predicative adjectives come after the noun to which they are linked with a copula or another linking mechanism. Finally, when the noun is elided, the adjective can play the role of head of a Noun Phrase (NP). In this case, it is called a nominal adjective (see example [3.4]).

The <u>blue</u> sea is beautiful.	Prepositive attributive adjective	
The governor – <u>general</u> decided to leave the country.	Postpositive attributive adjective	[3.4]
The sea is <u>blue</u>.	Predicative	
The customers prefer the <u>best</u> and the <u>cheapest</u>.	Nominal	

Verbs are morphologically variable in number, person, modality, time and voice. Preceded by their subjects and followed by their objects in the preferential order of English, they are the core of both verbal groups and sentences. Verb complements take on highly variable functions and forms, such as direct object complement, indirect object complement, infinitive and gerund. Several typologies exist for classifying verbs on the basis of their relationships with their complements: intransitive verbs, direct transitive verbs, indirect transitive verbs, verbs that take two complements, etc. Likewise, we will consider the relationship between the verb and its subject; defined subject: *he walks, John works*, indefinite subject: *somebody walks* or in which the subject is the agent of the action, as with the verbs *beat, attack* and *give* or in which it is the patient, as in *die*.

As emphasized by [CRE 95], the category of adverbs includes the most syntactically heterogeneous words; in other words, this category contains groups of words that cannot be categorized otherwise. These include words such as circumstantial adverbs including *somewhere now* and *tomorrow*; and adverbs of place such as *here, there* and *far*; of manner such as *slowly, softly* and *loudly*; and of degree, such as *very* and *too*. Generally invariant in English, we have some cases where two adverbs differ in number such as somtime and sometimes. However, in French, in unusual cases, the adverbs agree in number with the names or adjectives they modify as in example [3.5].

Les voies de l'avenir sont toutes/ grandes ouvertes devant Julien.	The paths of the future are all + plural/ widely + plural open in front of Julien. [3.5]
Les enfants derniers-nés.	The new + plural born babies.

The role of adverbs is to modify, particularly adjectives and verbs. One of the specific characteristics of many adverbs is that they play roles considered to be secondary; for this reason, they can be removed without causing a significant semantic change.

Determiners, which occur together with nouns, may vary in number, gender (like in French) and even person in the case of possessives. These elements are adjacent to nouns and can only be separated from them by attributive adjectives, sometimes preceded by one or more adverbs. Only one determiner can be used per noun, except in the case of quantifiers such as *all* and *other*. In French, there are cases where the gender or number of a noun cannot be determined, so determiners help solve the ambiguity, as in *un vieux* (an old) versus *des vieux* (det_plural old)*, un élève* (det_masc student) versus *une élève* (det_fem student), etc. This ambiguity is even more significant in spoken language since the same spoken form of some nouns can correspond to the written forms of the feminine singular, masculine singular or feminine plural. Note that the category of determiners includes several subcategories, including articles, possessive adjectives, demonstrative adjectives, indefinite adjectives, numeral adjectives and interrogative adjectives.

Pronouns act as substitutes for noun phrases. For this reason, they are highly variable in terms of gender, number and person. In terms of morphology, they can correspond to words such as *he, me* or *nothing*. This category includes a multitude of subcategories including demonstrative pronouns, possessive pronouns, relative pronouns, interrogative pronouns, indefinite pronouns and numeral pronouns (e.g. *two have left their homes*).

Prepositions are invariable and play an important role in the structuring of linguistic elements. As we have already seen, they serve to create certain compound words such as *on top of* and also introduce noun or indirect object complements as in example [3.6].

I would like to go <u>from</u> Toronto <u>to</u> Istanbul.   [3.6]

Johan Gave a gift <u>to</u> a stranger.

3.1.4. *Terms, collocations and colligations*

In this section, we will look at the different types of word clusters that are constructed on the basis of syntactic and/or semantic criteria. The recognition and correct processing of these clusters is an important factor in the understanding of any text. These clusters take various forms such as *technical* terms, collocations and colligations.

What we mean by "terms" includes a heterogeneous number of linguistic phenomena including common nouns, named entities and noun phrases that may be composed of other noun phrases. The difference between an ordinary word and a term is that the latter identifies a specific concept in a specialized field, such as *bovine spongiform encephalopathy* in the field of medicine. From a linguistic point of view, terms constitute a cohesive unit both syntactically and semantically, even if the term cannot stand as a sentence on its own.

Expressions such as *an awful lot, a good deal, kick the bucket* and *curriculum vitae* are particular linguistic units requiring specific treatment at the lexical, syntactic and semantic levels, and which are commonly known as collocations. Coined by Firth [FIR 57], the term *collocation* designates a group of words with strong statistical relationships. The weight attributed to the role of grammar in this relationship has been the subject of intense debate within the linguistic community. Traditionally, the terms phraseological units, idiomatic expressions, fixed expressions [MIS 87, GRO 96] and semi-fixed expressions [LAM 05] have been preferred. A more recent view considers words to have a linguistic charge, and to exercise an attraction or repulsion for other words [REN 07]. We refer readers to the summaries by [PAR 98] and [PEC 04] for a more complete overview. Note that the concept of collocation is closely linked to the analysis of corpora, which supports the calculation of co-occurrence statistics (see Table 3.5 for some examples in French literature).

Collocation		Balzac	Hugo	Zola
Faire fortune	Get rich	131.31	–	21.38
Faire faillite	Going bankrupt	74.96	14.58	23.27
Faire banqueroute	Going bankrupt	7.99	5.98	5.91

Table 3.5. *Examples of collocations in three French literary corpora [LEG 12]*

Another type of lexical cluster that should be mentioned is *colligation*. This is a multidimensional unit in which relationships of lexical co-occurrence have the same importance as the pattern of grammatical relationships of words [HUN 01, HOE 04] (see Table 3.6).

Pattern	Examples
The most + adjective	The most advanced The most rich
N/NP + replaced + by	The old car is replaced by a new one.

Table 3.6. *Examples of colligation*

Even if the use of such patterns is not tightly integrated within syntactic theories, these constructions have been the subject of various applications in the field of task-oriented dialog systems. Moreover, collocations and colligations are used in applications aimed at the classification of texts according to author or era.

3.2. Automatic morphological analysis

Modern computers have enough memory to store all the inflected forms of a language. However, devices with limited memory such as mobile phones and handheld computers, require more compact methods for storing these various forms. Likewise, certain languages, as we have seen, have a very rich morphology that makes their raw data very heavy to store. For a general introduction to the questions of computational morphology, we refer readers to [ROA 07] and [BEE 03].

3.2.1. *Stemming*

Steming is the simplest form of morphological processing. It consists of returning inflected, derived or compound words to a canonical form, called the lemma, which is not necessarily a word or morpheme in the language. Generally speaking, the goal behind lemmatization is to identify the concept or approximate meaning of the word. Used particularly in the field of information retrieval, lemmatization gives more flexibility to textual information retrieval by making it possible to find words that are morphologically different but conceptually similar. For example, the words *consolatory* and *constancy* are stemmed into the lemmas *consolatori* and *constanc* respectively that are not part of the English vocabulary. Likewise, in French, the lemma *mainten*, which is neither a word nor a morpheme, corresponds to the words *maintenir* (keep), *maintenant* (now), *maintenait* (have kept), *maintenaient* (have kept in plural), *maintint* (kept), etc.

Several approaches have been proposed for automatic lemmatization. Some are more statistical, while others are more heuristic or based more or less on linguistic criteria.

3.2.1.1. *Successor variety*

The successor variety approach was proposed by Margaret Hafer [HAF 74]. Influenced by structural linguistics, this approach attempts to identify the boundaries between morphemes based on the distribution of phonemes as observed in a large corpus. The basic criterion is the number of letters likely to follow the current chain in the corpus. To illustrate the function of this approach, imagine that we have a micro-corpus containing the following words: able, ape, beatable, fixable, read, readable, reading, reads, red, rope and ripe. To find the value of the successor for a word such as *read*, we create a table of successors (see Table 3.7).

Prefix	Successors	Letters
R	3	E, I, O
RE	2	A, D
REA	1	D
READ	3	A, I, S

Table 3.7. *Successors of the word read [FRA 92]*

With a larger corpus, the number of successors shrinks until we reach the boundary of a segment, and then increases. The question at this stage is one of establishing the boundary of the lemma. The simplest (and most controversial) solution consists of establishing a theoretical value for the number of successors. Since there are no concrete criteria for choosing this threshold, error is fairly likely. One heuristic approach consists of considering a boundary to have been reached when the number of occurrences of a letter is greater than that of the two letters that precede and follow it, respectively. Statistical criteria such as entropy have also been used (see [MAN 99]).

3.2.1.2. *Porter stemming algorithm*

Developed in the 1970s as part of an information retrieval project at Cambridge University in the United Kingdom, the Porter stemming algorithm is distinguished by its simplicity and effectiveness [POR 80], making it one of the most popular algorithms for lemmatization [WIL 06a]. Versions for several European languages has been proposed by [POR 06]. However, a detailed presentation and analysis of the algorithm is beyond the scope of our objectives in this book. Consequently, we will limit ourselves to a general description of its features.

In the first stage, we tag letters that correspond to vowels as v. A sequence of consonants or vowels with length greater than zero will be denoted by C and V, respectively. This yields the following general pattern: [C](VC)m[V], where the square brackets denote the arbitrary presence of their content and m is a constant that can be zero or more. Then a series of rules of the form (condition) S1 − > S2 are applied to remove the suffixes. The general meaning of these rules is as follows: S1 is replaced by S2 if the word is ending with the suffix S1, and the stem before S1 satisfies the given condition. Note that the condition may be expressed in terms of the value of m, such as m > 1. The rules are organized into groups that are applied in cascade. Some examples of transformations after this stage are given as follows:

SSES → SS	caresses → caress
(m>0) EED → EE	feed → feed
(m>0) ATIONAL → ATE	Relational → relate
(m>0) ICATE → IC	triplicate → triplic
(m>1) IZE →	bowdlerize → bowdler
(m>1) E →	probate → probat

3.2.1.3. *N-gram-based clustering*

This approach aims at the clustering of morphologically similar words without identifying the lemma. From a functional point of view, in many applications, this goes back to the same thing since the objective of the lemmatization process is to reduce the variety of morphological forms without the modular use of higher processing related to the syntax or semantics of the text.

The principle consists of identifying the n-grams of each word in the corpus and then calculating their similarity to other words using the Sørensen–Dice coefficient (equation [3.7]) [DIC 45]:

$$\text{sim} = \frac{2Z}{X+Y}$$ [3.7]

where:

– Z is the number of unique bigrams shared by both words;

– X is the number of unique bigrams in the first word;

– Y is the number of unique bigrams in the second word.

Take the French words *bonbon* and *bonbonne* (carboy). First, we will identify the bigrams that make up each of these two words. We must also identify the bigrams whose occurrence in the word is greater than one (see Table 3.8). Note that a trigram- (or higher) based approach is also possible, but the principle is the same.

Word	Bigrams	Repeated bigrams
bonbon	bo on nb bo on	bo on
bonbonne	bo on nb bo on ne	bo on

Table 3.8. *Bigrams of the words bonbon and bonbonne*

As we can see in Table 3.8, the word *bonbon* has five bigrams, two of which are repeated and the word *bonbonne* has six bigrams, two of which are also repeated. For our example, this gives: sim = 2*1/1+2 = 0.66. In a real application with a corpus of *n* words, we would then move on to calculating the distances between all of the words and storing the result of this calculation in a matrix of *n* rows and *n–1* columns. The matrix thus constructed serves to support a clustering algorithm used to find morphologically similar words.

3.2.2. *Regular expressions for morphological analysis*

First developed by Stephen Kleene [KLE 56], regular expressions are aimed at specifying classes of character chains through the specification of patterns. A chain is a sequence of symbols: numbers, letters, spaces, tabulations and punctuation marks. In this context, spaces are symbols like any other and can be represented by a specific chosen character. The simplest form of regular expressions is a simple character chain. Regular expressions are conventionally placed between two slashes / /. For example, the expression /Ouagadougou/ is used to recognize all chains containing the subchain *Ouagadougou*, such as "Ouagadougou *is an African capital*" or "*Ouagadougou is a city*". Some other examples are given in Table 3.9.

Expression	Recognized chains
/Ouagadougou/	Ouagadougou is the capital city of Burkina Faso.
/city/	The city of Damascus was among the most splendid cities in the world.
/play/	He played football during his entire life.

Table 3.9. *Some regular expressions with simple sequences*

Square brackets can be used to recognize a class of characters that is a set of equivalent characters. Look at the examples in Table 3.10.

Expression	Recognized category	Examples
/[Oo]uagadougou/	Ouagadougou ouagadougou	Ouagadougou is a modern city. Ouagadougou is far from here.
/[xyz]/	x, y or z	Lexique, zanini
/[0123456789]/	0, 1 or 2 etc.	un ensemble de 1 jusqu'à 9

Table 3.10. *Regular expressions with character categories*

We can use shortcuts to designate continuous sequences of characters as follows:

– [a–z] one character among all lower-case letters;

– [e–h] one character among e, f, g, h;

– [A–Z] one character among all upper-case letters;

– [0–9] a single digit.

We use special operators like Kleene operators, anchors, joker operators and disjunction to increase the power of regular expressions.

The Kleene star operator*, represented by the symbol *, means zero or more occurrences of the character or pattern that precedes it. For example, /b*/ recognizes a chain of zero or more *b* and /vv*/ recognizes a chain of one or more *v*. Likewise, the expression /[0–9]*/ recognizes any number, such as 124, 6547 or 098, as well as empty chains. The operator ? is a particular case of the operator * because it designates the preceding character or expression, or nothing. Thus, the expression /cars?/ is used to recognize the chain *cars* or *car*.

The Kleene operator + is used to render the occurrence of the previous character or regular expression mandatory at least once. In other words, in order to be satisfied, this operator requires one or more occurrences of what precedes it. Thus, the expression /goo+d/ is used to recognize *good* or *goood* or *gooood*, etc.

To express more specific limitations regarding the number of occurrences of any item, we use two curly brackets {*n*}, with *n* corresponding to the number of repetitions of the element. For example, /hello{5} the world/ recognizes *Hellooooo the world* but not *Hellooo the world*. A variant of this operator follows the syntax {n, m}, with *n* and *m* expressing the minimum and maximum limits, respectively, of a range of occurrences. For example, the expression /cour{2,4}/ will recognize *courr*, *courrr*, and *courrrr*, but not *cour* or *courrrrrrrr*. If the upper limit in the expression {n} is not specified, so it will recognize at least *n* occurrences of the preceding chain. Thus, the expression /Voila{3}/ recognizes *voilaaa* and *voilaaaaaa* but not *voila* or *voilaa*.

As its name suggests, the Joker operator, designated by a period, is a character that recognizes all other characters except the back slash \, which is a character that plays an important role in the syntax of regular expressions. For example, the expression /fin.r/ recognizes the chains *finar*, *finir*, *finur*, *finmr*, *finxr*, etc. Likewise, the expression /see.*future/ recognizes chains that begin and end, respectively, with *see* and *future*, such as *see the future* and *see the beginning of new future*.

Anchors constitute a set of operators, with various uses, having to do with the position of the regular expression in the chain. Here are some examples:

– ^ tags the starting character of the row;

– $ marks the end of the row;

– \b end of a word;

– \B not the end of a word.

For example, the expression /^The animals hunt.$/ specifies a row that contains only the chain *The animals hunt*. / \b44/ this regular expression recognizes the chain 44 in *Mary is 44 year old today* because it follows the end of the word.

The copy operator is used to memorize part of the chain in order to reproduce it identically later on. Suppose that we are trying to create an expression in which we have two identical subchains such as: the world is beautiful the world is crazy. To do this, we can tag the first chain using the operator () and the second using \1. The (.*) is beautiful. The \1 is crazy. Of course, we can repeat the copying to get new subchains, such as: the world is crazy. The (.*) is beautiful. The \1 is superb. The \2 is vaste.

The disjunction operator, which takes the form of a transverse bar, |, is used to express an alternative between two possibilities. Thus, the expression /flower | rose/ recognizes either the chain *flower* or the chain *rose.*

To combine multiple operators within one expression, we must be aware of their precedence. It is important to know that simple character chains have higher precedence than the disjunction character, which has lower precedence than all other operators. A list of operators ranked by priority is given in Table 3.11.

Operators	Precedence	Examples
Parentheses	Highest	()
Counters/repetition		* + ? {}
Sequences and anchors		^ $ \b \B
Disjunction	Lowest		

Table 3.11. *Priority of operators in regular expressions*

To understand the concept of precedence properly, let us take two examples. The verb *connaître* (to know) can be written in two ways in French: *connaître* and *connaitre*. To express this fact with a regular expression, all we need to do is use the disjunction operator to choose between everything to its left and everything to its right. This gives the following expression: connaitre | connaître. A more elegant way of expressing this divergence of writing consists of using two parentheses as follows: conna(i|î)tre. Since parentheses have higher priority than disjunction, disjunction is limited to sequences of characters between parentheses. Moreover, because repetition operators or counters, have higher priority than sequence operators, the expression hello* can interpret *hello* or *helloo*, while the expression (hello)* can interpret *hello hello* or *hello hello hello* due to the higher priority of parentheses.

Regular expressions have been used as a means of optimizing the process of creating morphological analyzers, as well as executing multiple word-retrieval tasks in a given corpus. For example, a regular expression can be used to find all the words in a corpus that share the same affix or to specify rules of morphological analysis. They have also been used for the syntactic analysis of temporal expressions such as dates of events and tokenization [KAR 97].

3.2.3. *Informal introduction to finite-state machines*

First introduced in 1945 by Warren McCulloch, a neurophysiologist, and Walter Pitts, a logician, finite-state machines (FSM) were developed to model the activity of a neuron that gives an output of 1 when active and an output of 0 when inactive [MCC 43]. According to this approach, each neuron receives input from the other neurons to which it is connected and possesses an activation function. Between 1951 and 1956, Kleene continued this work by defining finite-state machines and regular expressions and by proving the equivalence between them. Today, FSM are used in a large number of fields including NLP, particularly for morphological and syntactic analysis and the modeling of sequential logic circuits and communication protocols. For an introduction to FSM in the context of NLP in general, and morphology in particular, see [ROC 96, KAR 05, KAP 97] and [SIL 15].

Informally speaking, an FSM consists of a finite number of states and transitions. It is known that Markov chains, which we saw in Chapter 2, are formally equivalent to FSM, with the probability of transition on the arcs of a Markov chain being the only difference. Likewise, FSM are formally equivalent to regular expressions, which means that any regular expression can be expressed in the form of an FSM, and *vice versa*.

Let us begin by examining an FSM to express or recognize an expression of encouragement or surprise that can have the following forms: *well!*, *bravo!*, *very good!*, *very very good!*, *very very very good!*, etc. To do this, we need an FSM that includes four states and appropriate transitions between these states. Using a graph is an effective way to represent an FSM (see Figure 3.1).

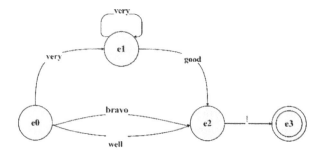

Figure 3.1. *FSM for expressions of encouragement*

As we can see in Figure 3.1, we have two arcs to go from the initial state e0 to the state e2 with two different words: *bravo* and *well*. Another important point is the cyclical transition to state e_1, which makes it possible to repeat the word *very* an unlimited number of times. Finally, the final state e_3 is marked with a double circle to indicate that it is a valid stopping point. This prevents the recognition of an incomplete chain such as *very* or *very good* (without an exclamation point) as a valid chain. Another way of representing the information contained in an FSM consists of constructing a so-called transition table. For our example, it takes the form of Table 3.12.

Input state	very	bravo	well	good	!
e0	e1	e2	e2	φ	φ
e1	e1	φ	φ	e2	φ
e2	φ	φ	φ	φ	e3
e3	φ	φ	φ	φ	φ

Table 3.12. *FSM transition table for expressions of encouragement*

FSM are equivalent to a type of grammar called regular grammar or type-three grammar (in Chomsky's hierarchy of formal grammars), which we will examine in Chapter 4. Likewise, we have seen that Kleene's work proved the equivalence of regular expressions to FSA. A conversion algorithm has even been suggested to automate the conversion of a regular expression into a machine [THO 68]. To illustrate the relationship between FSM and regular expressions, let us look at the examples of basic cases shown in Figure 3.2.

As we can see in Figure 3.2, the repetition of an element by the * operator is obtained via a cyclical link. Likewise, the step backward with an empty element represented by the symbol ε enables the infinite repetition of the sequence *ab* in the expression: (ab)+. Finally, the disjunction operator in the third expression is translated via an alternative route, starting from the same state.

In the context of NLP, it is important to distinguish between two types of FSM: deterministic machines and non-deterministic machines, which can be considered as different formal variants for expressing the same thing. It has been shown that for each non-deterministic FSM there is a deterministic equivalent that recognizes exactly the same language (see [HOP 01] for a discussion of the proof).

Regular   FSM
expression

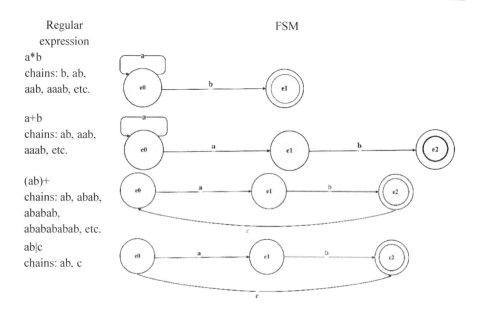

a*b
chains: b, ab,
aab, aaab, etc.

a+b
chains: ab, aab,
aaab, etc.

(ab)+
chains: ab, abab,
ababab,
ababababab, etc.

ab|c
chains: ab, c

Figure 3.2. *Examples of regular expressions with their FSM equivalence*

A deterministic FSM is a machine which, with a given input chain, always carries out the same processing. This means that each transition in a deterministic FSM always leads to one single state for a given symbol being processed. For example, the FSM that corresponds to the expression a*b in Figure 3.2 is a deterministic FSM. As we can see, there is always a single destination state no matter what symbol is analyzed.

On the other hand, a non-deterministic FSM is not required to obey the constraint of the uniqueness of analysis for a given chain and may, therefore, allow a multitude of destination states from the same initial state and with analysis of the same symbol. There are generally two reasons for the non-determinism of a machine: empty transition and cyclical transition. The FSM of the expression a+b in Figure 3.2 is an example of a non-deterministic machine with a cyclical transition. We can see that from the initial state e_0 and with the symbol *a* being analyzed, we can reach either state e_0 with the cyclical transition, or state e_1 with the other transition.

From a formal point of view, an FSA can be defined as a quintuplet: $<E,$ $\Sigma, \Delta, e_0, F>$, where:

– E is a finite number of states;

– Σ is a finite number of input symbols;

– Δ is a function used to find the next state. This function gives a single state for a given symbol in the case of a deterministic machine and can give multiple states in the case of a non-deterministic machine;

– $e_0 \in E$ is the initial state;

– $F \subseteq E$ is the group of final states. In exceptional cases, all the states of a machine can be final states.

The intuitive application of FSM to morphological analysis consists of dedicating one transition between two states for each morpheme; for example, in processing the French verbs *poser* (put) and *porter* (carry), which can take one of sixteen possible prefixes as well as a fairly large number of suffixes depending on time and modality. To avoid repetitions, we can design an FSM like the one shown in Figure 3.3, in which these prefixes are divided into three subgroups; two groups for the prefixes proper to each verb and one group for the prefixes shared between them. An empty transition is used to pass shared prefixes to the stems of the two verbs concerned. For the sake of simplicity, we have shown present indicative tense suffixes only.

A more refined way of representing information consists of considering phonemic level rather than morphemic level. This makes it possible also to take into account the phonological phenomena discussed in the previous chapter. For example, this increased fineness allows us to insert the vowel *e* at the end of the stem of the French verb *manger* (eat) with the plural first-person present indicative suffix *–ons*. This, of course, entails an additional cost in terms of the time needed to prepare such a machine and makes the use of an automatic compilation module to generate it indispensable.

Despite its interest, a machine like the one shown in Figure 3.3 allows us to see whether a word conforms to the morphological rules of a given language without giving any more information about its internal structure. To do this, we need more advanced tools, such as finite-state transducers (FST), which we will examine in the next section.

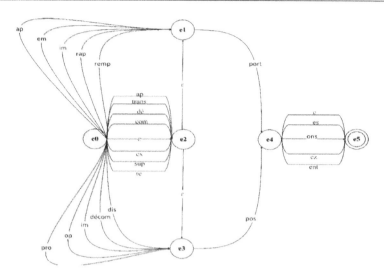

Figure 3.3. *Conjugation of the verbs poser and porter in the present indicative tense*

3.2.4. *Two-level morphology and FST*

The origins of two-level morphology lie in the work of Noam Chomsky and Morris Hall, conducted in the 1960s on the SPE model we saw in Chapter 2. It is an ordered sequence of rules of rewriting that convert abstract phonological representations into surface representations via a series of intermediary representations (see [ANT 91a, ANT 91b, KOS 83], and [KAR 01] for an introduction). The pioneers of FST-based morphological analysis, namely Lauri Kartunen, Martin Kay, Ronald Kaplan and Kimmo Koskenniemi (the four Ks), met at a conference at the University of Texas at Austin. Koskenniemi discovered the idea of FST reinvented by Kay and Kaplan and subsequently developed it for morphological and morphophonological analysis.

In 1972, C. Douglas Johnson published his thesis entitled *Formal Aspects of Phonological Description*, in which he showed that phonological rules are less powerful than they appear to be [DOU 72]. Johnson observed that the same rules of rewriting, which are dependent on context, can be applied recursively to their own output. Phonologists have observed that the point of application can be applied to the left or right of the chain after each application. For example, if the rule $x \rightarrow y \mathbin{/} z \_ w$ is used to rewrite the chain "uzxwv" as "uzywv", all applications following the same rule must leave the y unchanged and affect only the parts "uz" or "wv". Johnson showed that this property could be modeled by an FST, a result that was rediscovered by Ronald Kaplan and Martin Kay in 1981 [KAP 94].

In two-level morphology, words are represented by a correspondence between two levels, the lexical level and the surface level. The lexical level is a simple concatenation of morphological information describing the nature and properties of the word in question. These items of information are attached to one another via a group of features such as: +N (noun), +V (verb), +Pl (plural), +Sg (singular), +PART-PRES (present participle), +PAST-PART (past participle), 3SING (3rd person singular), +Masc (masculine), +Fem (feminine), etc. The surface level affects the graphic form of the word; that is, the sequence of its characters (e.g. *house*). In other words, two-level morphology represents a word as a series of complex sequences called *correspondence pairs*. Analysis is, thus, defined as the establishing of a relationship between the surface level and the lexical level. As an example, here are the correspondence pairs for some words:

– houses: H:H O:O U:U S:S E:E +N:ε +Pl:S;

– looks: L:L O:O O:O K:K +V:ε +3P:S +Sg:ε;

– choice: C:C H:H O:O I:I CC E:E +N:ε +Sg: ε.

As we can see in the above examples, each phoneme corresponds to a phoneme when it is part of a stem. For lexical morphemes reflecting one aspect of the totality of the stem such as +N and +V in order to indicate whether it is a noun and a verb, respectively. We have an empty equivalent at the surface level, represented by the symbol ε. Morphemes are marked on the surface by a single character such as *s*, which marks the plural of a noun, or *e*, which marks the ending of the verb at the end of the noun. This pair can be represented hierarchically as shown in Figure 3.4.

Figure 3.4. *Correspondence pair for the word houses*

Note that this method of representation is very well-suited for incorporating modifications at the surface level using the phonological constraints that we saw in Chapter 2, such as deletion and assimilation. Thus, in certain contexts, the stop consonant [d] is replaced by the silent occlusive [t]. This can be easily represented by the correspondence pair D:T E:E. A case of marked assimilation in French consists of pronouncing the word *cheveu* (hair) [ʃəvø] as [ʃfø], with the removal of the ə and the replacement of the consonant [v] with [f] under the effect of the silent consonant [ʃ]. With two-level morphology, this gives us the correspondence pair C:C H:H E:ε V:F E:E U:U.

According to the representation in Figure 3.4, analysis is viewed as a passage from the surface level to the lexical level, while the process of generation consists of moving in the opposite direction, from the lexical level to the surface level. It is clear that the FSM we saw in the previous section are not capable of executing both of these procedures. For this, researchers have suggested using Finite State Transducers (FST) which are finite-state machines annotated with two tags on each transition, where each tag corresponds to a pair of symbols (see Figure 3.5).

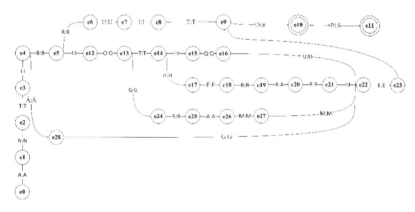

Figure 3.5. *FST for some words in French with the prefix "anti–"*

As we can see in Figure 3.5, words are represented in terms of characters rather than morphemes. This makes it possible to gain detailed knowledge of the phenomena described by transition pairs. Besides, this allows us to avoid repeating certain characters, particularly at the start and end of words, thus giving a more compact representation.

As we saw in section 3.1.2.3, derivation is another interesting phenomenon. To exemplify the way in which derivations are represented in

FST, we have chosen an example of some stems of which the derivations are shown in Figure 3.6. This FST shows how to move from a noun such as the French word *concept* to an adjective such as *conceptuel*, and then to a verb such as *conceptualiser* and finally on to a new noun with *conceptualisation*. For the sake of clarity, we have used the infinitive form of the verb only, but it is easy to imagine the complexity of the FST with all the possible forms for variations according to time and modality. The feminine and plural forms of adjectives, as well as the plurals of nouns, are also considered. This shows the flexibility given by an FST, which covers morphological aspects as well as phonological ones such as the doubling of the *l* in the feminine version of an adjective like *conceptuelle*.

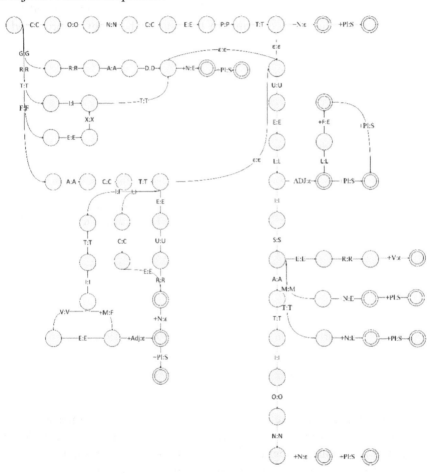

Figure 3.6. *Partial FST for the derivation of some French words*

In its role as recognizer, an FST takes a correspondence pair as input and yields as output a confirmation or negation of the matching of the surface level with the lexical level, knowing the language of the correspondence pairs defined by the FST. In its role as analyzer, the FST takes a surface chain as input and must produce the corresponding lexical chain. Conversely, when acting as a generator, the FST produces the surface chain from the lexical chain provided as input. Finally, the FST can act as a translator by taking a chain as input and producing a transformed version of this chain as output.

It is important to mention that rules of rewriting can be converted into transducers. In fact, each cascade of transducers can be converted into a single transducer which establishes a relationship between the lexical forms on one hand and the surface forms on the other. The diagram proposed by Kay and Kaplan is shown in Figure 3.7.

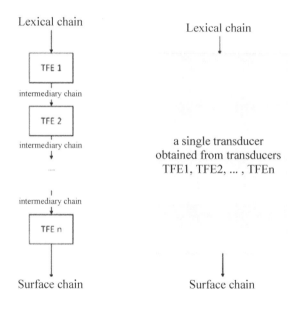

Figure 3.7. *Kay and Kaplan diagram*

In real applications, the rules and lexicon are first defined as regular expressions. These rules are then converted by a compiler into an FST. The compiler is a computer program designed to translate expressions expressed using any formal language, in this case regular expressions into expressions in another formal language (FST) (see Figure 3.8).

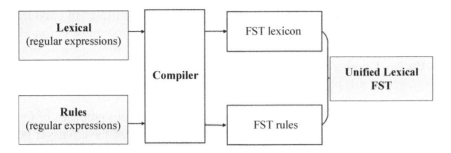

Figure 3.8. *Xerox approach to the use of FST in morphological analysis*

Lexical transducers can contain hundreds of thousands or even millions of states and arcs. The size of the FST is particularly large when morphologically rich languages such as German are involved.

3.2.5. *Part-of-speech tagging*

Part-of-speech (POS) tagging consists of automatically attaching a part-of-speech tag to all of the words in a given corpus. This process is very important in many NLP applications. In the context of syntactic analysis, POS tagging facilitates the task of the analysis module by providing it with a starting point. As part of information retrieval, the annotation of words in a given corpus may improve search functions by giving more weight to categories known to better characterize texts, such as nouns and adjectives, rather than categories judged to be less representative semantically, such as determiners, pronouns, verbs and adverbs. In the field of speech synthesis, we saw examples in Chapter 2 in which the grammatical category is used to select the appropriate phonetic transcription for two homographs in different grammatical categories. We have also covered examples like the word *live*, the pronunciation of which changes depending on whether it is noun or a verb. This interest explains the multitude of approaches proposed by many research centers all over the world. For an introduction to this issue, we refer readers to [VOU 09] and [ASM 14].

The main obstacle that must be overcome by a POS tagging system is ambiguity. There are many words, or more precisely graphic forms, in languages such as French and English, which can be associated with multiple parts of speech. Some examples are *dogs* (verb, noun), *heat* (verb, noun), slow (adjective, verb, adverb), *in* (preposition, adverb, adjective, noun), *large* (adjective, noun, adverb), etc. In a study on English conducted by Steven DeRose, it was demonstrated that, though ambiguity is found in only 11.5% of categories, in reality, it concerns around 40% of the words in the Brown Corpus, the subject of the study [DER 88]. This shows the importance of this phenomenon and the necessity of an effective solution.

In addition to ambiguity, modules must overcome various problems of form. These problems include the processing of new words that have not yet been observed in the training corpus. This problem, which is common to several NLP applications, makes it necessary to use contextual and morphological heuristics among others. Another problem observed in corpora of real data has to do with spelling errors and various types of grammatical errors. In transcribed spoken language corpora, this can be manifested in the form of hesitations, repetitions, false starts, etc. This phenomenon is difficult to model and causes changes of context, making it difficult to assess the part of speech of the current word. The writing systems of certain languages, such as Arabic, can also be an additional source of ambiguity that can be qualified as artificial. In this language, the diacritical marks corresponding to short vowels are usually optional. For example, without diacritical marks, the grapheme كتب can correspond to different words such as [kataba] *wrote* and [kutub] *books*.

There are lists of tags called *tagsets* for several languages. Though true standards do not exist on this subject, some lists have been commonly adopted, such as the *Penn Treebank tagset*, which includes between 36 and 41 tags depending on the version. This tagset is based on the one developed for the annotation of the Brown corpus [NEL 64, GRE 81, MAR 93]. The Xerox tagger[1] has 77 categories for English, 45 for French and 67 for German. As an example, we have provided a list of eleven basic tags drawn from the *Universal Part-of-Speech Tagset* in [BIR 09] (see Table 3.13). The definition of lists like these is only part of a wider convention that must specify, among other things, the way in which compound and conflated words must be tagged. For example, the *Penn Treebank* uses the same tag for each part of the word. Another possibility consists of joining the components of a compound word with an underscore to indicate a single word. Likewise, conflated words require a special convention. It is common to split them when possible into multiple parts according to rules such as *I'll = I will*. In cases where morphemes cannot be separated, a possible way to process them consists of a replacement like in *gonna = going to*.

---

1 https://open.xerox.com/Services/fst-nlp-tools/Consume/Part%20of%20Speech%20Taggin20%28Standard%29-178.

Tag	Category	Examples
ADJ	Adjective	beautiful, legal, cold, round
ADP	Preposition	from, at, to, with
ADV	Adverb	slowly, far, here, now
CONJ	Conjunction	and, or, if, but, whereas
DET	Determiner	the, a, some
NOM	Noun	house, book, table, France, Delaware
NUM	Numeral	one, thirty-five, 2016, 22.5
PRON	Pronoun	he, she, we, they, them
VERB	Verb	eat, see, live
AUX	Auxiliary	is, was, has
.	Punctuation	.; ,: ?
X	other	dunno, ersatz

Table 3.13. *A minimal list of tags*

Using the tag list shown in Table 3.13, we can tag the micro-corpus shown in Figure 3.9.

*the/DET   first/ADJ   written/**ADJ**   records/NOUN   for/ADP   the/DET history/NOUN of/ADP France/NOUN was/VERB in/ADP the/DET Iron/ADJ Age/NOUN*

what/PRON is/VERB now/ADV France/NOUN made/VERB up/PRT the/DET bulk/NOUN of/ADP the/DET region/NOUN known/ADJ to/ADP the/DET Romans/NOUN as/ADV Gaul/NOUN

Roman/ADJ   written/**ADJ**   resources/NOUN   noted/VERB   the/DET presence/NOUN   of/ADP   three/DET   main/ADJ   ethno-linguistic/ADJ groups/NOUN in/ADP the/DET area/NOUN

*the/DET   Gauls/NOUN   history/NOUN   is/AUX   written/**VERB**   long/ADJ time/NOUN ago/ADV*

Figure 3.9. *A micro-text tagged with POS*

In terms of linguistics, part-of-speech tagging is done on the basis of multiple sources of information. First, at the lexical level, we need a database in which all of the graphemes with their possible tags are stored. Next, we need contextual information that is both syntactic and semantic. Morphology plays a significant role as well. As we have seen, the suffixes of words formed with derivation make it possible, in certain cases, to figure out the part of speech to which the word belongs. For example, it is possible to assume with a reasonable degree of certainty that a word ending with –*tion* is a noun. However, because many suffixes are shared by multiple parts of speech, this is not always a decisive source of information. In a tagging module, it is common to use low-level sources of information, avoiding higher-level information like semantics. This is due partly to the fact that it is possible to get good results with low-level information, and partly to the dependency of the task and the heaviness of development that such higher-level information causes.

The tagging process is, generally, carried out in two main stages: identification of potential tags for input words and disambiguation. The first stage consists of using a dictionary in which the words with their possible tags are stored. The second, disambiguation stage consists of choosing the most appropriate tag among the candidates identified in the first stage.

In terms of technologies, a fairly large number of approaches have been used to construct taggers. Statistical approaches and learning-based approaches have frequently been used to solve problems of ambiguity in the domain of NLP. These approaches are good for modeling problems that are still poorly understood in theoretical terms. This has motivated a considerable amount of research since the 1980s [DER 88, GAR 87, CHU 88, SCH 94b]. Several variants of learning-based approaches such as memory-based learning [DAE 96, DAE 10], neural networks [SCH 94a, MAR 96] and genetic algorithms [ARA 02, BEN 13] have also been tested with varying degrees of success.

3.2.5.1. *Statistical approaches*

The simplest statistical approach consists of using n-grams, the simplest form of which, as we saw in Chapter 2, are unigrams. We intuitively count frequencies of categories per word in an annotated corpus. During automatic tagging, we choose the most probable category for this word, independent of its context. Suppose that we have a tagger trained on the corpus shown in

Figure 3.9 and that this module must tag the sentence *The written history of the Gauls is known*. To do this, we must resolve the ambiguity of the word *written*, which, as we have seen, can be tagged with both the verb and adjective categories. Based on the statistics of our micro-corpus, the probability that *written* can be tagged with the category ADJ is 2/3 = 0.66, while the probability of its tagging with VERB is 1/3 = 0.33. Therefore, the category ADJ is chosen for this sentence. The same choice is made for the sentence *The history was written*, which is not the correct choice. Note that, depending on the language considered, the expansion of the contextual window by taking into account a larger value for *n* in an n-gram-based model does not always guarantee improved results. This is mainly due to the fact that the larger the history, the more data it requires. For example, a study conducted on three southeast Asian languages has shown that the results of tagging with unigrams are better than those with bigrams. However, the same study showed that tagging with hidden Markov models (HMMs) yields results that are clearly superior to n-grams [HAS 07]. To benefit the most from n-grams even with a limited amount of data, we can use a so-called *backoff heuristic*, which consists of tagging all the words we can tag with trigrams and leaving the rest for lower levels. The process is then repeated with bigrams and then unigrams, and finally unknown words are tagged with the most frequent category.

A more refined way of modeling the problem consists of considering HMMs as a framework [CHA 93a]. This uses two different complementary sources of information: the probabilities of transition from one tag to another $p(e_i|e_{i-1})$, which model the context of occurrence, and the probabilities of generation of a given word knowing a specific tag $p(W_i|E_i)$, which takes the lexical level into account. These two probabilities are calculated using equations [3.8]. We refer readers to the section on HMMs in Chapter 2 for further details.

$$p(W|E) = \prod_{i=1}^{n} p(m_i \mid e_i) \qquad\qquad [3.8]$$

$$p(E) = \prod_{i=1}^{n} p(e_i \mid e_{i-1})$$

As we saw in the speech sphere, we can make three types of inference with HMMs by using different algorithms. Firstly, in a context of language modeling, we can calculate the probability of a sequence of words with a hidden sequence of tags. The other possibility is to find the most probable tag sequences, knowing the sequence of words observed. Finally, we can estimate the parameters of the model based on the observed sequence and the corresponding tag sequence.

Let us return to our example: *The written history of the Gauls is known.* To analyze this sentence with an HMM, we can have the two representations below in the form of a Bayesian network, the first of which is correct. The probabilities on the arcs of the networks in Figure 3.10 are obviously calculated from the corpus being used.

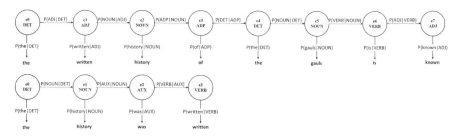

Figure 3.10. *Tag sequences for "The written history of the Gauls is known"*

The main problem with the statistical approaches we have just presented is the need for significant linguistic resources, which are not always available, particularly for minority languages or those from the countries of the south. Approaches based on Bayesian networks, the generic form of HMMs, have been proposed, among other solutions, as an unsupervised learning framework for tagging modules [LEE 10, GOL 07]. Though the results of these approaches are encouraging, this work is still in the research stage and its results are clearly inferior to those of supervised approaches.

3.2.5.2. *Transformation-based approach*

Proposed by Eric Brill at the University of Pennsylvania, the transformation-based approach is a hybrid process combining statistical information with symbolic rules [BRI 93]. According to this approach, the tagging process is carried out according to the architecture in Figure 3.11.

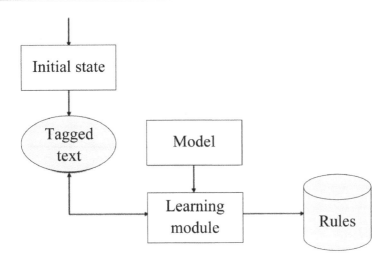

Figure 3.11. *Architecture of the Brill tagger [BRI 95]*

The first stage of processing consists of rough tagging using an n-gram based module (in the initial version, a unigram was used). Words tagged in this way and liable for improvement are marked for possible correction in a subsequent stage. The rules to be applied to the corpus being tagged are learned automatically from the corpus so as to guarantee maximum improvement. The process of applying a rule is called a transformation. There are two types of transformations; rules and triggers. Rules specify the way in which the input must be modified, while triggers serve to describe the conditions of application of a given rule.

Trigger	the previous word is in the Tag_z category
Rule	if *trigger* then $\text{Tag}_x \rightarrow \text{Tag}_y$

For example, we can have the following trigger and rule:

Trig_{01}	The previous word is in Category DET.
Rule	If Trig_{01} then VRB $\rightarrow$ NOM

Transformations are learned as follows. First, the training corpus is annotated with the initial tagger. The results of the annotation are compared with the annotation of this corpus and the number of errors is obtained. Then, we apply a number *n* of possible transformations, and for each transformation we calculate the number of errors. The transformation that reduces most of the errors is chosen. We repeat this process with the rest of the transformations until we arrive at a state in which there are no more transformations that might improve tagging. An example of learning adapted from Brill [BRI 95] with four possible transformations is given in Figure 3.12.

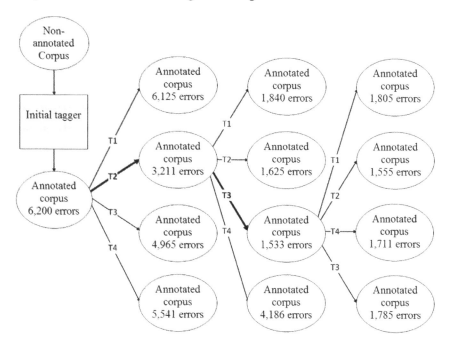

Figure 3.12. *Example of transformation-based learning*

As we can see in Figure 3.12, the transformation T2 is the one that gives the minimum number of errors; therefore, it is chosen as the first transformation to be carried out. Next, we choose T3 for the same reason. Finally, since none of the transformations improve the results, we stop our search.

This approach is distinguished not only by its effectiveness and good results, but also in terms of data savings. Unlike statistical approaches requiring the storage of n-gram models, the size of which increases along with the size of *n*, the Brill approach requires much less memory. This is particularly appreciated in mobile telephone systems or similar applications.

The final advantage of this approach concerns the readability of internal representations in the form of rules that can be understood by humans. Finally, it is probably worth to mention that a generic toolbox for the transformation-based learning algorithm is available [FLO 02] and transformation-based tagging has been applied to several languages.

Syntax Sphere

4.1. Basic syntactic concepts

4.1.1. *Delimitation of the field of syntax*

Syntax is a key branch of linguistics. It focuses on the scientific study of the structure of the sentence as an independent unit. The word order, the dependency relationships between these words and, in some languages, the relationships of agreement as well as the case marking, are among the points that attract the attention of most of the researchers. The final objective of syntax is to produce a formal description of underlying regularities with regard to sentence organization and to determine the principles that govern the combination and dependency relationships of words and word sequences within the sentence.

Syntax, which is actually at the heart of linguistics, maintains fairly close relationships with the other branches of linguistics, including phonology, morphology and semantics.

With phonology and more particularly with prosody, the relationships are well-known. For example, the syntactic process of emphasis, which manifests itself in the form of dislocation (Yesterday evening, John came to see me), or clefting (It is John who came to see me yesterday evening) is systematically accompanied by a particular intonation (see [NES 07, INK 90] for an introduction to these issues).

Compared to morphology, syntax is distinguished by the fact that it focuses on the relationships between words, whereas morphology focuses on variations of word forms. Note that some linguistic currents consider that the morpheme is the basic unit of syntax. This leads us to consider that the processes of word creation and sentence construction are of the same nature. So, in this case we refer to morphosyntax.

Essentially formal, syntax focuses on the linguistic form of the sentence without giving paramount importance to the meaning, which is the object of study for semantics. If we want to simplify, we can say that syntax focuses on the relationships between linguistic signs, while semantics focuses on the relationships between these signs and those signified by them, as well as on the overall meaning of the sentence that will be produced by syntax. However, the boundaries of the two disciplines are not very clear. In fact, it is being widely accepted that the complementarity of these two sources of knowledge is indispensable to be able to correctly understand a sentence, particularly in the case of syntactic ambiguities or semantic anomalies (see [ANT 94, MAH 95], for a review of these studies in the field of psycholinguistics and NLP).

4.1.2. *The concept of grammaticality*

As linguistics is a descriptive and non-normative discipline, it is appropriate to begin with a clarification of the descriptive concept. In the NLP field, grammar is not a set of rules that a speaker must follow as its production is considered to be well-formed (normative grammar), but rather a description of the syntactic phenomena used by any linguistic community at a given time. This description is, therefore, used as a reference to distinguish what is said from what is not said.

According to Chomsky, grammar is a device capable of carrying out grammaticality judgments, that is to classify input units sequences (lexical strings) in two groups: the correctly and the incorrectly formed strings (see Figure 4.1). It is this device which characterizes the competence of an average speaker.

Figure 4.1. *The role of grammar according to Chomsky*

This original conception of grammar has important implications on the concept of grammaticality. On one hand, grammaticality is different from the concept of frequency of use of a phenomenon within a linguistic community, made evident by the number of occurrence(s) of the phenomenon in question, in a corpus which is considered to be representative of the language of this community. Let us take a look at the sentence [4.1]:

The astronaut's .... telescope is broken.   [4.1]

The dotted lines in the sentence [4.1] can be replaced by several words that would be both syntactically and semantically acceptable, including *spatial*, *infrared* and *black*. Furthermore, although the words *table* (noun) and *temporal* (adjective) are statistically unlikely in this context, only the adjective *temporal* allows us to create a grammatical sentence:

    a) The astronaut's spatial telescope is broken.

    b) The astronaut's temporal telescope is broken.   [4.2]

    c) The astronaut's table telescope is broken.*

The group of sentences [4.2] leads us to the second important distinction between grammaticality and interpretability. In fact, sentence (a) is interpretable, both grammatically and semantically, whereas sentence (b) is syntactically acceptable but not semantically interpretable. Finally, sentence (c) is neither grammatically nor semantically interpretable.

Note that grammaticality is not a necessary condition for comprehension. Although agrammatical, the famous sentence: *Me Tarzan, You Jane*, is quite understandable.

4.1.3. *Syntactic constituents*

Parsing consists of the decomposition of sentences in major syntactic units and of the identification of dependency relationships. This analysis often leads to a graphical representation in the form of a box according to the original approach proposed by the American linguist Charles Francis Hockett [HOC 58]. Generally, it takes the form of a tree diagram. As we will see later, this type of analysis has a strong power of explanation and particularly takes into account the syntactic ambiguities. But the question arises as to what is the nature of the constituents of a sentence. Is a constituent a word or a word sequence which has a particular syntactic role within the sentence?

This is the question which we will try to answer.

4.1.3.1. *Words*

As we have seen in the sphere of words, in spite of the problems related to its definition, the word is recognized more or less explicitly as a syntactical unit by different theories from different currents. In addition, many NLP applications presuppose a linguistic material in which words are labeled. That is why we believe that it is useful to begin with a classification of words according to their syntactic categories, commonly known as parts of speech.

Some linguists deny the existence of linguistic units which are higher than words. That is why they assume that syntactic relationships are limited to the dependencies between the words in the sentence. In the tradition of the Slavic language, the French linguist Lucien Tesnière has laid the foundations of a linguistic theory known as dependency grammar [TES 59]. Recovered, developed and applied in the works of several linguists throughout the world [HAY 60, MEL 88, HDU 84, HUD 00, SLE 91], dependency grammar is still a minority current in modern syntax.

From a formal point of view, a dependency tree (or a stemma), according to Tesnière's terminology, is a graph or a tree that represents the syntactic dependencies between the words of a sentence. Thus, dependency grammar is based on these explicit- or implicit-dependency relationships rather than on a precise theoretical framework. The shapes of the trees vary according to specific theories in spite of shared theoretical foundations.

For example, the arguments of a verb (subject, object, adverbials, etc.) are all syntactic functions, represented by arcs, which come from the verb. In the grammatical framework of *Word Grammar* (WG) by Hudson, we have four relationships: pre-adjoint, post-adjoint, subject and complement (see Figure 4.2).

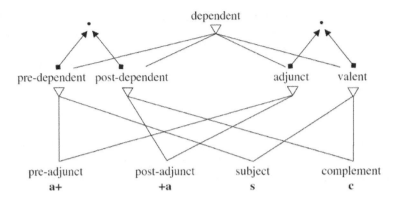

Figure 4.2. *Relationships in the framework of formalism, WG [HUD 10]*

Like other dependency formalisms, WG bestows a central place to the verb which has the pivotal role in the entire sentence (see example in Figure 4.3).

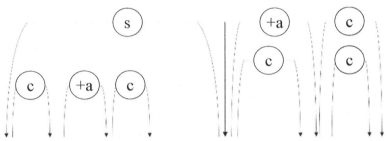

The linguistics professor gives lectures at the university

Figure 4.3. *Analysis of a simple sentence by the formalism of WG*

The adoption of the word as the central unit of syntactic analysis has several advantages. In fact, on one hand, it is easier to establish relationships with phonological levels, such as binding phenomena and semantics with semantic roles. On the other hand, it is particularly suitable for processing phenomena such as discontinuity which are, among others, observed in German [KRO 01]. As for the disadvantages, we can note the difficulty of processing phenomena such as coordination which often involves word blocks and not simple words.

A discussion of the formal equivalence of dependency grammars with the formalisms based on superior constituents is proposed in [KAH 12]. We should also note that parsing modules based on dependency grammars have been developed for several languages including English [SLE 91], French [CAN 10] and Arabic [MAR 13].

4.1.3.2. Clauses

The concept of *clause* is transdisciplinary to the extent that it is used in a variety of disciplines such as linguistics and logic. In the field of logic, it is a utterance which accepts a truth value: it can be either true or false.

In the field of linguistics, it refers to sequences of words containing at least a subject and a verb predicate explicitly or implicitly present (in the case of an ellipsis). A sentence can, generally, be divided into several clauses each contained in the other. A more detailed discussion of these issues follows later in section 4.1.3.6.

4.1.3.3. Phrase

A *phrase*, according to classical terminology, is a word or a consecutive word sequence which has a specific syntactic role and which we can, consequently, associate with a single category. This unit is defined based on the criterion of substitutability. It is, thus, located among the higher unit of syntax, i.e. the sentence and the word. However, some believe that this concept, as relevant as it is, does not justify by itself that we devote to it a full-level of analysis, because the information that a phrase conveys is already present at the level of the words that comprise it (see the previous section).

Each phrase has a kernel of which it inherits the category and the function. A kernel can be a simple word or a compound; it can even be another phrase. Sometimes, a phrase has two or several kernels which are either coordinated or juxtaposed [4.3]:

My friend Michael and his best friend Jack go to the
factory every morning.   [4.3]

A human being eats, drinks and breathes all the time.

One of the specificities of the phrase is that it is a recursive unit. In other words, a phrase can have as a unit another phrase as in [4.4], where the prepositional phrase (of the village) are included in the principal phrase, the doctor of the village:

The <u>doctor of the village</u> left last night.   [4.4]

The question that now arises is: how can we identify a phrase with the help of rigorous linguistic tests? In fact, several experiments or linguistic tests allow us to highlight the unitary character of a word sequence, such as commutation, ellipsis, coordination, shift, topicalization, clefting and negation.

Commutation or substitution is the simplest test which consists of replacing a series of words by a single word without changing the meaning or the grammaticality. Pronominalization is the most well-known form of this process as in [4.5]:

<u>My good cousin Philip/He</u> wants to go to Sweden.
                                                        [4.5]
I would like this red notebook and *that white notebook/that one.*

Ellipsis is one of these tests, as it allows us to substitute a word sequence with a zero element. In French, noun phrases can be elided, unless they are subjects or obligatory complements, prepositional phrases, as well as phrasal complements (see [4.6]):

a) Cynthia will <u>go at her office,</u> and her husband will go as well.
                                                        [4.6]
b) This teacher <u>will send a bouquet of flowers</u> and Mary as well.

As we can see in [4.6], we replace only the complement noun phrase of an indirect object with a null element (a) and we also replace the verb phrase and its complement with a null element (b).

In the case of discursive ellipses, responses to partial questions can be achieved with an ellipsis. In the group of responses [4.7], case (a) represents the possibility of a response without omitting the subject and the object noun phrases. Response (b) provides a possible response with only the complement noun phrase. Response (c) gives us a case of a truncated constituent, while response (d) gives us an example of a response with a very long and, therefore, not acceptable sequence as a constituent.

– What do you want?

– I would like a linguistics book.

– a phonetics book. [4.7]

– a phonetics book tonight (it is not a constituent).

– * phonetics (it is not a constituent).

Coordination can also serve as a test to identify units, because the basic principle of coordination is to coordinate only the elements of the same category and whose result is an element which keeps the category of coordinated elements. Let us look at the examples [4.8]:

a) Np [Np [My math teacher] and Np [my next door neighbor]]
will leave together.

b) Julia Vp [Vp [loves the country] and Vp [hates the big cities]].

c) S [S [I prepared dinner] and S [I vacuumed]]. [4.8]

d) S [S [You should leave immediately] or S [you will miss your exam]].

e) S [S [I searched for the books] but S [I did not find them]].

As we can see in example [4.8], we can coordinate noun phrases (a), verb phrases (b) or entire sentences (c, d and e).

Dislocations or shifts consist of changing the location of a set of words that form a constituent. There is a variety of shift types, such as topicalization, clefting, pseudo-clefting, interrogative movement, shift of the most important constituent to the right.

Topicalization is the dislocation of a constituent at the head of the sentence by means of a separator such as a comma:

a) I love phonetics.

b) Phonetics, I love it.

c) I can write a poem. [4.9]

d) To write a poem, that I can.

In the series of example [4.9], we have a shift of the noun phrase *phonetics* in (b) while a shift of the verb phrase *write a poem* in (d).

Interrogative movement is also a common case of the interrogative that is to separate the complement noun phrase from the object as in [4.10]:

- I want a Syrian restaurant in the city center.
- which Syrian restaurant in the city center do you prefer?

[4.10]

Clefting consists of highlighting a constituent of the sentence *CONST* by delimiting it, respectively, by a presentative and a relative. Two main types of clefting exist in French: clefting on the subject and clefting on the object. To these, we can also add pseudo-clefting. The patterns that follow these three types are presented in Table 4.1.

Type	Pattern	Examples
Clefting on the subject	It is CONST. that Y.	It is <u>John</u> that will go to the market. It is <u>the neighborhood postman</u> that brings the letters every day.
Clefting on the object	It is X that CONST.	It is <u>carrots</u> that John will buy from the market tomorrow evening. It is <u>tomorrow evening</u> that John will buy carrots from the market.
Pseudo-clefting	What X has done, it is CONST.	What he asked the school principal, it is <u>a question</u>. To whom he asked a question, it is <u>the journalist from Washington Post</u>. To whom he asked a question, it is <u>to the school director</u>.

Table 4.1. *Clefting patterns*

Restrictive negation (which is sometimes called exceptive) can also be used as a test, because only a constituent can be the focus of the restrictive negation. In fact, from a semantics point of view, it is not truly a negation but rather a restriction that excludes from its scope everything that follows. As we can see in Table 4.2, these excluded elements are constituents which have varied roles.

Example	Excluded constituent
The newly arrived worker says only <u>nonsense</u>.	NP/Direct object complement
I am only <u>a poor man</u>.	NP/Subject attribute
She eats only <u>in the evening with her best friends</u>.	NP/Adverbial phrase
She eats in the evening only <u>with her best friends</u>.	PP/Indirect object complement
It is only the young from Toulon that <u>says nonsense</u>.	VP
He only <u>says nonsense</u>.	VP

Table 4.2. *Examples of restrictive negation*

Note that sometimes restrictive negation excludes more than one constituent at a time as in [4.9] even if, in this case, we can consider the relative act as a genitive construction:

> Il *n'*y a *que* ta thèse qui t'intéresse.
> (It is only your thesis that interests me.)   [4.11]

Another test consists of separating a constituent from its neighbor by using the adverbs *only* or *even* whose role is to draw the line between the two constituents. Let us look at the examples [4.12]:

> a) *Only/Even* <u>Frank</u> studies at night his syntax exercises.
>
> b) Frank studies <u>only/even at night his syntax exercises</u>.   [4.12]
>
> c) Frank studies at night <u>only/even his syntax exercises</u>.

As we note in group [4.12], this test does not allow us to identify only a single boundary of a constituent. Consequently, when the delimited element is at the head of the sentence as in (a). On the contrary, in one of the contexts as in the case of (b), we must perform other tests for complete identification.

4.1.3.4. Chunks

Proposed by Steven Abney, these are the smallest word sequences to which we can associate a category as a nominal or verb phrase [ABN 91a]. Unlike phrases, chunks or segments are non-recursive units (they must not have a constituent of the same nature). That is why some as [TRO 09] prefer to call them *nuclear phrases* or *kernel phrases*. Just like phrases, chunks typically have a keyword (the head) which is surrounded by a constellation of satellite words (functional). For example, in *the tree in the meadow*, there are, in fact, two separate chunks: *the tree* and *in the meadow*. A more comprehensive example of the analysis is provided in Figure 4.4.

The effort to establish such a conclusion of course must have two foci, the study of the rocks and the study of the sun.
{[Det the] } { [N effort] } { [Inf-To to] [S to] } { [V establish] } { [Predet such] [Det such] [Pron such] } { [Det a] } { [N conclusion] } { [Adv of course] }{[N must] [V must] } { [V have] } { [Num two] } { [N foci] } {[Comma ,]} . . .
[DP [Det the] [NP [N effort]]]
[PC-Inf [IP-Inf [Inf-To to] [VP [V establish]]]]
[DP [Predet such] [Det a] [NP [N conclusion]]]
[PC [IP [AdvP [Adv of course]] [Modal will] [VP [V have]]]]
[DP [NP [Num two] [N foci]]]
[Comma ,]
[DP [Det the] [NP [N study]]]
[PP [S of] [DP [Det the] [NP [N rocks]]]]
....

Figure 4.4. *Example of an analysis by chunks [ABN 91a]*

Although it does not offer a fundamentally different conception on the theoretical level, the adoption of chunks as a unit of analysis has two advantages. On the one hand, it allows us to identify more easily the prosodic and syntactic parallels, because the concept of a "chunk" is also prosodically anchored. On the other hand, the simplification of syntactic units has opened new roads in the field of the robust parsing (see section 4.4.10).

4.1.3.5. *Construction*

Unlike other linguistic formalisms which establish a clear boundary between morphemes and constituents of higher rank as a syntactical unit, the construction grammar (CG) argues that the two can coexist within the same theoretical framework [FIL 88] (see [GOL 03, YAN 03] for an introduction). According to this concept, grammar is seen as a network of construction families. The basic constructions in this CG are very close to those of the HPSG formalism that we are going to present in detail in section 4.3.2.

4.1.3.6. *Sentence*

As the objective of syntax is to study the structure of the sentence of which it is the privileged unit, it seems important to understand the structure of this unit. The simplest definition of a sentence is based on spelling criteria according to which it is a word sequence that begins with a capital letter and ends with a full stop. From a syntactic point of view, sentence decomposes in phrases, typically a noun phrase and a verb phrase for a simple sentence. Similarly, it consists of a single clause in the case of a simple sentence or of several clauses in the case of a complex sentence. We distinguish between several types of complex sentences based on the nature of the relationships that link the clauses. These are the following three: coordination, juxtaposition and subordination.

Coordination consists of the comparison of at least two clauses within a sentence by means of a coordinating conjunction, such as *and, or, but, neither*, etc. (see [4.13]):

– It is early *and* minors are already moving towards the entrance.
– John loves to observe the underwater life *but* he cannot swim.
– Joana will study a Master degree in Charlottesville *or* she will   [4.13]

   work in a bank in Lynchburg.

Juxtaposition consists of using two joint clauses by a punctuation mark that does not mark the end of a sentence [4.14]. It is, according to some, a particular case of coordination to the extent that we can, in most cases, replace the comma with a coordinating conjunction without changing the nature of the syntactic or semantic relationships of the clauses:

It snows a lot in Delaware this winter, truck
drivers will have many problems.   [4.14]

The relationship of subordination implies a relationship of domination between a main clause which serves as the framework in the sentence and a dependent clause which is called subordinate. Often linked by a complementizer which can be a subordinating conjunction (that, when, as, etc.) or a relative pronoun (who, what, when, if, etc.), the subordinating clauses are sometimes juxtaposed without the presence of a subordinating element that connects them as in [4.15]:

The more he eats, the more he gains weight.   [4.15]

Traditionally, we distinguish between two types of subordinates, namely the completives [4.16] and the relatives [4.16] and [4.16]:

a) Cedric believes that all software should be free.

b) Cedric estimates the distance between the two cities.

c) Last night I met John *who* was your neighbor in Brooklyn.   [4.16]

d) Celine sings all songs *that* she finds interesting.

In the case of completives, the subordinate plays the role of a completive to the verb of the main clause. Thus, the completive in [4.16a] has the same role as the complement to the direct object of [4.16b]. With regard to relatives, they complement a noun phrase in the main clause that we call antecedent. For example, *John* and *all songs* are, respectively, the antecedents of relative subordinate clauses in [4.16c] and [4.16d].

4.1.4. *Syntactic typology of topology and agreement*

Topology concerns the order in which the words are arranged within the sentence. In general, topology allows us to know the function of an argument according to its position in relation to the verb [LAZ 94]. For example, French is a language with the order SVO (subject-verb-object). Other languages are of the SOV type, such as German, Japanese, Armenian, Turkish and Urdu, whereas languages such as Arabic and Hawaiian are of the VSO type. Note that, depending on the language, this order can vary from fixed to totally variable. We will refer to [MUL 08] for a more complete presentation of the word order in French.

The relationship of agreement consists of a morphological change that affects a given word due to its dependency on another word. They are a reflection of the privileged syntactic links that exist between the constituents of the sentence. In French, it is a mechanism according to which a given noun or pronoun exerts a formal constraint on the pronouns which represent it, on the verbs of which it is a subject, on the adjectives or past participles which relate to it [DUB 94].

4.1.5. *Syntactic ambiguity*

Syntactic ambiguity, which is sometimes called amphibology, concerns the possibility to associate at least two different analyses with a single sentence. Unlike lexical ambiguities, the source of the syntactic ambiguity is not the polysemy of the words that make up the sentence, but rather the differences of dependency relationships that the constituents of the sentence can have. Therefore, we refer to attachment ambiguity of which the most notable case is the attachment of the prepositional phrase see Figure 4.5 for an example.

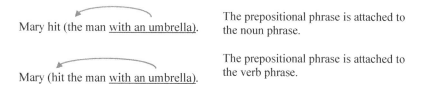

Mary hit (the man <u>with an umbrella</u>).   The prepositional phrase is attached to the noun phrase.

Mary (hit the man <u>with an umbrella</u>).   The prepositional phrase is attached to the verb phrase.

Figure 4.5. *Example of attachment ambiguity of a prepositional phrase*

The attachment of the adjective to the nominal group can sometimes cause an ambiguity of syntactic analysis as in [4.17]:

a) (red birds) and (fish).

b) Red (birds and fish).

c) a wheel of a (car used for...).

d) (a wheel of a car) used for.

[4.17]

Coordination consists of coordinating two or several elements that have the same syntactic nature. In this situation, there may be an ambiguity when there are two coordinating conjunctions of which we cannot delimit the scope. For example, in the sentence [4.18], we have two conjunctions *and* and *or* without knowing whether each of these conjunctions focuses on a simple element (tea, coffee) or on the result of the other conjunction: (sugar and coffee) and (tea or coffee):

a) I want (sugar and tea) or coffee.

b) I want sugar and (tea or coffee).

[4.18]

Ambiguity can also focus on the attachment of *the adverb* as in the French examples [4.19]:

a) il veut (bien apprendre). [He wants to (learn well).]

b) (il veut bien) apprendre. [(He wants well) to learn.]

[4.19]

In group [4.19], in case (a) the adverb *bien* (well) depends on the verb *apprendre* (learn), whereas in case (b) this same adverb depends on the semi-auxiliary veut (wants) with which it also forms a construction of collocation type.

Obviously, syntactic ambiguity directly affects the semantic interpretation of a sentence. In fact, sometimes, multiple analyses that allow the syntax are all semantically interpretable as in the sentences [4.19] and [4.17]; and sometimes there is only a subset of these analyses that is semantically interpretable as in the sentence [4.20]:

a) A (drug for atherosclerosis expired).

b) A drug (for atherosclerosis expired).

[4.20]

We note in the group [4.20] that only structure (a) where the adjective *expired* qualifies the noun *drug* can receive a semantic interpretation, whereas (b) where the adjective *expired* qualifies the noun *atherosclerosis* does not have a particular meaning. With this example, we return to the concept of grammaticality which was discussed at the beginning of this section.

4.1.6. *Syntactic specificities of spontaneous oral language*

In the field of speech, quite a considerable number of studies has focused on phonetic and phonological aspects. However, syntax, which is a central discipline in linguistics, is the only one to remain subject to the reign of the scripturocentrism as highlighted in [KER 96]. In fact, syntactic studies have focused primarily on the written word while neglecting the oral dimension, which was considered as an impoverished and sometimes deviant form of writing. The lack of linguistic resources due to the difficulties of collecting and transcribing spoken dialogues (see [BLA 87] for a general review of these issues), as well as the relatively limited importance of the syntactic processing of the spoken language before the 1990s are some other reasons for this delay.

4.1.6.1. *Topology in spoken language*

Spoken language does not seem to obey the same standard as the written text regarding word order. For example, the utterances of Table 4.3 are perfectly possible in a spoken conversation or in a pseudo-writing used in discussion forums on the Internet, while they are not acceptable in a standard written text:

Example	Structure	Order of elements
My notebook I forgot it at home	Anteposition of an NP	OSV
At 200 meters you will find a pharmacy	Anteposition of an PP	OSVO
Me my father I love him very much	Double marking	SOSVO

Table 4.3. *A few examples of variation of the word order at the oral framework*

The question that arises is to know what is the importance of these cases in terms of frequency in the spoken conversations and then to know whether this frequency depends on the syntactic context (i.e. is it more important in a syntactic context C1 than in another syntactic context C2?). [ANT 01] have tried to answer these questions in their study which is based on three corpora of spoken French[1]. Thus, these researchers have shown that in ordinary situations the finalized language respects the privileged sequencing.

---

1 These are the three corpora: Air France [MOR 92], Murol [BES 95], Ozkan [OZK 94].

4.1.6.2. *Agreement in gender and number*

According to the constructions, agreement is often respected in the spoken language, but not always. For example, non-respect of the agreement between the noun and/or its adjectives is very rare, while non-agreement in gender between the attribute and the word to which it relates is very frequent [4.21] [SAU 72]:

a) Une voiture émetteur   A car (+fem) emitter (+masc)

b) Les revenus salariaux   The wage (+fem) incomes (+masc)   [4.21]

c) Les dispositions que   The provisions (+fem) that we have
nous avons pris   taken (+masc)

4.1.6.3. *Extragrammaticalities of the oral language*

Extragrammaticalities, unexpected structural elements, spontaneities and disfluencies are among other terms that have been proposed in the literature to designate the spontaneous phenomena of the spoken language such as hesitation, repetition, self-correction, etc. [LIC 94, SHR 94, HEE 97, COR 99, MCK 98]. Each of these terms has its motivation. The term "extragrammaticality" which has been adopted by [CAR 83] seems to be the most appropriate, because it is sufficiently general and specific to cover the different spontaneous phenomena of the spoken language which do not depend directly on the syntax of the language. Three key phenomena are distinguished within extragrammaticalities: repetitions, self-corrections and false starts.

Repetition is the repetition of a word or a series of words. It is defined on purely morphological criteria. Consequently, the formulation and the paraphrase of a utterance or a segment (where we repeat two segments that have the same meaning) are not considered as repetitions: *it would be a Paris Delhi flight rather than a flight a domestic flight*.

Repetition is not always a redundancy. It can also have a communicative function. For example, when a speaker is not sure if his message (or a part of his message) will be perceived clearly by its audience because of a poor articulation, a noise in the channel, etc., he repeats it. In addition, repetition is a quite common pragmatic means to mark an affirmation or an insistence as in the utterance [4.22]:

yes yes I'll buy one of these   [4.22]

In the utterance 4.22, the repetition of the word *yes* has an affirmative function.

Self-correction is to replace a word or a series of words with others in order to modify or correct the meaning of the utterance. Self-correction is not completely random and often focuses on a segment that can have one or several phrases [COR 99]. That is why it is frequently accompanied by a partial repetition of the corrected segment. Let us look at the utterance [4.23]:

Yes : <u>I have</u> *a* <u>I have</u> *the* web pages yes.   [4.23]

In this utterance, self-correction is performed by repeating the segment *I have* and by replacing the word *a* with the word *the*. We note that the two words have the same morphological category (definite article) and the same syntactic function (determiner).

A false start is to abandon what has been said and starting over with another utterance. Syntactically, this is manifested with the succession of an incomplete (or an incorrectly formed) segment and with a complete segment. Let us look at the utterance [4.24]:

(…) yes <u>*it is at*</u> and this is taken to the second floor.   [4.24]

Unlike self-correction, there is no analogy between the replaced segment and the rest of the utterance. Thus, we can notice in the example [4.24] that the abandoned segment *it is at* has almost no relationship with *this is taken*…. This form of extragrammaticality is the most difficult to process given that the detection criteria (essentially the incompleteness of a segment) are very vague and can lead to many problems both of overgeneration and of undergeneration.

4.2. Elements of formal syntax

4.2.1. *Syntax trees and rewrite rules*

After reviewing the different constituents of a sentence and the subtleties of the relationships that may exist between these constituents, we will now address the question of the analysis of the sentence by assembling the pieces of the puzzle. To do this, we will begin with phrases and end with complex sentences.

The idea to use rules to describe the syntax of a particular language goes back to the beginning of the past century. It was formalized in the 1950s, particularly with [CHO 56, BAC 59].

A syntax tree is a non-oriented (there is not a predetermined direction to traverse the tree) and acyclic (we cannot traverse the tree and then return to the starting point) graph consisting of nodes connected by arcs. A node represents a constituent or a morphological or syntactic category connected by an arc with the dominant node. Each constituent must be directly dominated by the corresponding category. Inspired by family relationships, the dominant node is called "parent node" and the dominated node is called "child node". Similarly, the highest node of a tree (the one that dominates all other nodes) is called the "root node" of this tree, whereas the lowest nodes in the hierarchy and which, therefore, do not dominate other nodes are called "leaf nodes". In syntax trees, each word is dominated by its morphological category and phrases by their syntactic categories. Logically, the S (Sentence) is the root node of these trees and the words of the analyzed sentence are the leaf nodes.

Let us begin with the noun phrases. As we have seen, a noun phrase can consist of a proper noun only in a common noun surrounded by a wide range of varied satellite words. Let us look at the phrases in Table 4.4.

Noun phrase	Sequence of categories
John	NP
My small red car	Det Adj N Adj
The house of the family	Det N prep Det N

Table 4.4. *Examples of noun phrases and their morphological sequences*

As we can see in Table 4.4, each noun phrase is provided with the sequence of morphological categories of which it is composed. This sequence provides quite important information in order to understand the syntactic structure of these phrases, but it is, however, not sufficient because it omits the dependency relationships between these categories. To fill this lack of information, syntax trees can shed light on both the order of constituents and their hierarchy. Let us examine the syntax trees of the phrases provided in Figure 4.6.

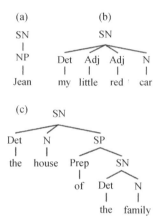

Figure 4.6. *Syntax trees of some noun phrases*

As we note in Figure 4.6, all the trees have an NP as a root node (the highest node) since they all correspond to noun phrases. Tree (a) is the simplest one since it is a proper noun which is capable of achieving a noun phrase without another constituent. In tree (b), the kernel of the phrase, the noun *car*, is qualified by an anteposed adjective *small* and by a postponed adjective *red*. Tree (c) shows how a prepositional phrase acts as a genitive construction. This prepositional phrase consists in turn of a preposition and a noun phrase.

A rewrite rule is an equation between symbols of two types: terminal and non-terminal symbols. Terminal symbols are symbols that cannot be replaced by other symbols and which correspond to words or morphemes of the language in question. Non-terminal symbols are symbols that can be replaced by terminal symbols or other non-terminal symbols. In grammar, they correspond to morphological categories such as noun, verb and adjective or to syntactic categories such as NP, VP, PP, etc.

The transformation of the syntax trees shown in Figure 4.6 in rewrite rules provides the grammar in Figure 4.7.

$$
\begin{array}{lll}
\text{NP} & \rightarrow & \text{PN} \\
\text{NP} & \rightarrow & \text{Det (Adj) N (Adj)} \\
\text{NP} & \rightarrow & \text{Det N PP} \\
\text{PP} & \rightarrow & \text{Prep NP} \\
\text{Det} & \rightarrow & \text{the} \mid \text{my} \\
\text{NP} & \rightarrow & \text{John} \\
\text{N} & \rightarrow & \text{car} \mid \text{house} \mid \text{family} \\
\text{Adj} & \rightarrow & \text{small} \mid \text{red} \\
\text{Prep} & \rightarrow & \text{of}
\end{array}
$$

Figure 4.7. *Grammar for the structures as shown in Figure 4.6*

Let us look at the noun phrase of the Figure 4.8. It is a noun phrase which includes an adverb and an adjective. What is specific in the phrase is that the adverb does not modify the noun but rather the adjective which in turn qualifies the noun. To understand this dependency relationship, the creation of an adjective phrase seems to be necessary.

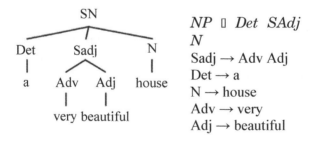

Figure 4.8. *Syntax trees and rewrite rules of an adjective phrase*

One of the specificities of natural languages is the production capacity of an infinite number of sentences. Among the sources of this generativity is the ability to emphatically repeat the same element especially in the spoken language. Let us look at the sentence [4.25], where we can repeat the adjective an indefinite number of times:

A flower beautiful beautiful …   [4.25]

This poses a problem because we have to repeat the same rule each time with an increase in the number of symbols (see grammar in Figure 4.9).

```
NP → Det N Adj
NP → Det N Adj Adj
NP → Det N Adj Adj Adj
etc.
```

Figure 4.9. *Grammar for the structures presented in Figure 4.8*

An elegant and practical solution to this problem is to use recursive rules that contain the same non-terminal symbol, both in its left and right hand sides (grammar in Figure 4.10).

```
NP → Det N SAdj
SAdj → Adj SAdj
SAdj → Adj
```

Figure 4.10. *Grammar for the noun phrase with a recursion*

According to the first rule of the mini grammar in Figure 4.10, a noun phrase consists of a determiner, a noun and an adjective phrase. According to our second rule, an adjective phrase consists of an indefinite number of adjectives: it adds an adjective and is called the rule of the adjective phrase. According to the third rule, the adjective phrase can consist of a single adjective and has the function to stop the looping to infinity of the second rule.

The adverbial phrase has for a kernel an adverb which can be modified by another adverb as in [4.26]:

very quickly

too late

[4.26]

Note that in order to avoid circularity in the analysis, we must distinguish the adverbs of degree (very, little, too, etc.) from other adverbs in the rewrite rules. This provides the rules of the form: AdvP → *AdvDeg* Adv.

The verb phrase is, as we have seen, the kernel of the sentence because it is the bridge that connects the subject with potential complements. Let us look at the sentences [4.27] with different complements:

I have seen the rose. (Direct Object)

He leaves tomorrow. (Adverbial phrase/Adverb)

Nadine gave a flower to her father kindly.

(Direct Object + Indirect Object)

[4.27]

These three sentences are analyzed with the syntax tree of the Figure 4.11.

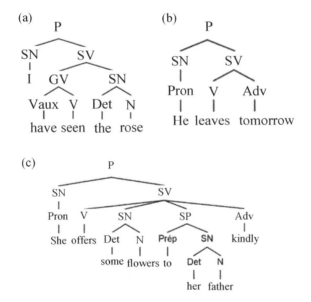

Figure 4.11. *Examples of VP with different complement types*

Complex sentences is another type of syntactic phenomena which deserves to be examined in the framework of syntactic grammar. What is significant at this level is to be able to represent the dependencies of the clauses within the sentence.

Completives consist of adding a subordinate which acts as a complement in relation to the verb of the main clause. Thus, the overall structure of the sentence remains the same regardless of the type of the complement. The question is: how should this phrasal complement be represented? To answer this question, we can imagine that the complex sentence has a phrase of a particular type (SPh (Sentence Phrase)) which begins with a complementizer (comp) followed by the sentence. Figure 4.12 shows a parallel between a phrasal complement and a pronominal complement (ordinary NP).

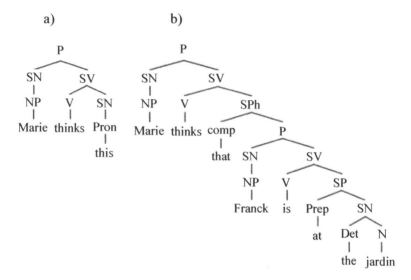

Figure 4.12. *Analysis of two types of sentences with two types of complements*

The processing of relatives is similar to the processing of cleft sentences to the extent that we consider the subordinates as phrasal complements. Naturally, in the case of relative subordinates, the attachment is performed at the noun phrase as in Figure 4.13.

Coordinated sentences consist of two or several clauses (simple sentences) connected with a coordinating conjunction. The analysis of this type of structures is quite simple since it implies a symmetry of the coordinated constituents dominated by an element that has the same category as the coordinated elements. This rule applies both at the level of constituents and at the level of the entire sentence (see Figure 4.14).

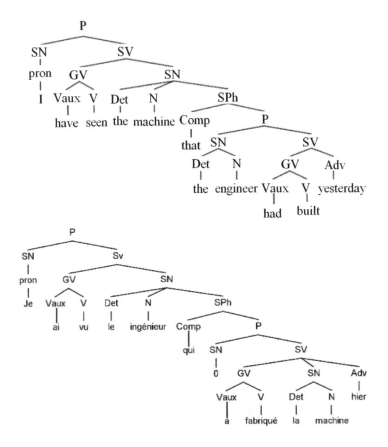

Figure 4.13. *Example of analysis of two relative sentences*

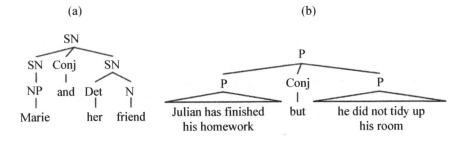

Figure 4.14. *Examples of the coordination of two phrases and two sentences*

We should also note that syntax trees are a very good way of highlighting the syntactical ambiguity which is manifested by the allocation of at least two valid syntax trees for the same syntactic unit (see Figure 4.15).

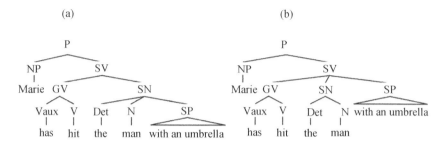

Figure 4.15. *Two syntax tree for a syntactically ambiguous sentence*

At the end of this section, it seems necessary to note that phrase structure grammars, despite their simplicity and efficiency, are not a perfect solution to understand all syntactic phenomena. In fact, some linguistic cases pose serious problems for the phrase structure model [4.28]:

– Caroline is too shy, I am not. (ellipsis)

– Frank, my colleague, a good pharmacist,
  died yesterday. (juxtaposition)   [4.28]

– Me, my house, its roof, it is damaged. (clefting)

– The toys, the child broke all of them. (interrelationships)

4.2.2. *Languages and formal grammars*

Formal language is a set of symbol strings of finite length, constructed on the basis of a given vocabulary (alphabet) and which is sometimes constrained by rules that are specific to this language (see [WEH 97, XAV 05] for a detailed introduction). Vocabulary, which is conventionally represented by lowercase letters of the Latin alphabet, corresponds to the words of the language which are the produced strings. To describe a formal language, the simplest way is to list all the strings produced by this language. For example, $L_1 = \{a, ab, ba, b\}$. The problem is that formal languages often produce an infinite number of strings whose listing is impossible. This requires the use of a formulation which characterizes the strings without having to list them all.

Let us look at L_2, a formal language which includes all non-null sequences of the symbol a: L_2 = {a, aa, aaa, aaaa, ...}. As it is impossible to list all the words of the language, we can use an expression of the form: {a^i | $i \geq 1$} where no limit is imposed on the maximum value of i, so as to be capable of generating an infinite number of strings. For a linguistic initiation to phrase structure and formal grammars, we refer to the books by Lélia Picabia and Maurice Gross [PIC 75, GRO 12].

When we refer to formal language, it is intuitive to evoke the concept of formal grammar. In fact, a formal language can be seen as the product of a grammar that describes it. Formally, such a grammar is defined by a quadruplet: **G** = (V_N, V_T, S, S) where:

– **V_N**: the non-terminal vocabulary;

– **V_T**: this vocabulary brings together all of the terminals of the grammar, which are commonly called the *words* of the language;

– **P**: the set of the rewrite rules of grammar (production rules);

– **S**: sometimes called an axiom, it is a special element of the set **V_N** which corresponds to well-formed sentences.

Note that the sets V_N and V_T are disjointed (their intersection is zero) and that their union forms the vocabulary V of the grammar. In addition, V* denotes the set of all the strings of finite length which are formed by the concatenation of elements taken from the vocabulary including the null string and V+ is equal to V* except for the fact that it does not contain the empty string: V+ = V*– {Φ}.

Rewrite rules have the following form: $\alpha \rightarrow \beta$ where $\alpha \ni$ V+ (however, it contains a non-terminal element) $\beta \ni$ V *.

Conventionally, lowercase letters at the beginning of the Latin alphabet are used to represent the terminal elements *a*, *b*, *c*, etc. We also use lowercase letters at the end of the Latin alphabet to represent strings of terminal elements *w*, *x*, *z*, etc. and the uppercase letters *A*, *E*, *C*, etc. to represent non-terminal elements. Finally, lowercase Greek letters represent the strings of terminal and non-terminal elements α, β, γ, δ, etc.

Let us consider a formal grammar $G = (V_N, V_T, P, S)$ with:

$V_N = \{S\}$

$V_T = \{a, b\}$

P: 1. S →aS

   2. S → b

The language defined by this grammar has the following form: L (G) = a*b. This means that the strings which are acceptable by this language are formed by a sequence of zero or more occurrences of *a* followed by a single occurrence of *b*: *b*, *ab*, *aab*, *aaab*, *aaaab*, etc.

To generate these language strings from the grammar, we begin with the special symbol *S*. We apply the set of rules which have the letter "S" in their left-hand side until there are no more terminals in the input string. The process to obtain strings from a grammar is called derivation.

Thus, in our grammar, the simplest derivation is to apply the rule #2 to replace the symbol *S* with the terminal b: S => b. Similarly, we can derive the sentence *aab* by the successive application of the rule #1 twice and the rule #2 only once: S => aS => aaS => aab.

4.2.3. *Hierarchy of languages (Chomsky–Schützenberger)*

After having presented languages and formal grammars, we can rightfully ask the following questions. What is the expressive power of a particular grammar? In other words, can a grammar G_1 describe all the languages produced by a grammar G_2? Can it describe other languages? How can we decide whether two languages described by two different formalisms are formally equivalent? To answer all these questions, the American linguist Noam Chomsky and the French mathematician Marcel Paul Schützenberger have proposed a framework for classifying the formal grammars and the languages they generate according to their complexity [CHO 56, CHO 63]. As this framework is capable of characterizing all recursively enumerable languages, it is commonly referred to as the hierarchy of languages or Chomsky hierarchy. It is a typology which includes four types of grammars, each included in the higher type and numbered between 0 and 4.

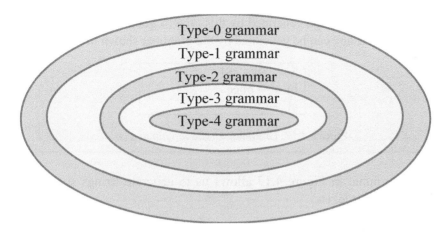

Figure 4.16. *Hierarchy of formal grammars*

Type-0 grammar, which is also called unrestricted grammar, is the most general form of the grammar because it allows us to transform an arbitrary non null number of symbols into an arbitrary number of symbols (potentially zero symbols). It refers to grammar which accepts, for example, the rules whose right-hand side is longer than their left-hand side, as the rule: A N E → a N.

Although they are the most general, these types of grammar are the least useful for linguists and computer scientists. The languages generated by a grammar of this type are called recursively enumerable languages and the tool to recognize them is the Turing machine[2].

Type-1 grammar, which is also called context-sensitive grammar, are types of grammars whose rules follow the following pattern: $\alpha A\beta \rightarrow \alpha\gamma\beta$.

With a non-terminal symbol A and sequences of terminal or non-terminal symbols α, γ and β knowing that α and β can be empty unlike γ. Another characteristic of these grammars is that they do not accept rules whose right-hand side is longer than the left-hand side.

2 The Turing machine is a hypothetical device capable of manipulating symbols on a tape. This concept was proposed in 1936 by the English mathematician Alan Turing.

The typical language generated by this type of grammars is of the form: $a^n b^n c^n$. This language can be generated by the grammar shown in Figure 4.17.

$$S \rightarrow abc \mid aSE$$
$$bEc \rightarrow bbcc$$
$$cE \rightarrow Ec$$

Figure 4.17. *Grammar for the language $a^n b^n c^n$*

The grammar in Figure 4.17 allows us to generate strings such as: abc, aabbcc, aaabbbccc, etc.

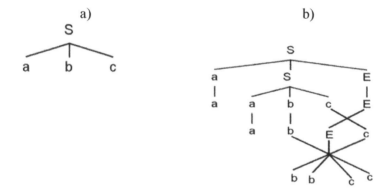

Figure 4.18. *Syntax tree for the strings: abc and aabbcc*

As we can see in Figure 4.18, the derivation of the abc string is performed in a direct way with a single rule: $S \rightarrow abc$, whereas the derivation of the aabbcc string requires contextual rules such as: $bEc \rightarrow bbcc$.

Type-1 grammar has been used for syntactic and morphological analysis particularly with augmented transition networks.

Type-2 grammar, which is called context-free grammar or context free grammar (CFG) is a type of grammar in which all rules follow the pattern: A $\rightarrow \gamma$. With a non-terminal symbol A and a sequence of terminal and non-terminal symbols γ. The typical language generated by Type-2 grammar is: $a^n b^n$ (see Figure 4.19 for an example).

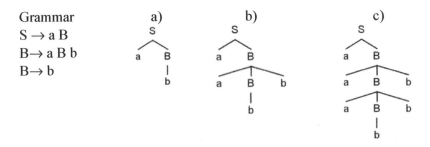

Figure 4.19. *The derivation of strings: ab, aabb, aaabbb*

To obtain the string *ab* in tree [4.19a], it is sufficient to apply the first rule: $S \rightarrow a B$ and then we replace "B" with "b" with the rule $B \rightarrow b$. In tree (b), we apply the rule: B $\rightarrow$ a B b only once, whereas in tree (c) we must apply it twice.

Thanks to its simplicity, Type-2 grammar is quite commonly used, particularly for parsing. The automaton which allows us to recognize this type is the recursive transition network (RTN). As we will see in section 4.4.2, there is a multitude of parsing algorithms of various types to perform parsing of natural languages with context-free grammar.

Several forms, which are called normal, have been proposed to simplify Type-2 grammar without reducing their generative capacities. Among these forms, those of Chomsky and those of Greibach deserve to be addressed. Note that these normal forms retain ambiguity. This means that an ambiguous sentence generated by a Type-2 G grammar is also ambiguous in languages G' and G'', which are the normal forms of this grammar according to Chomsky and Greibach formats, respectively. Finally, note that the equivalence between a CFG and its standardized form is low because, although both generate exactly the same language, they do not perform the analysis in the same way.

Chomsky's normal form imposes a clear separation between the derivation of terminal elements and the derivation of non-terminal elements. Thus, a grammar is said to be in Chomsky normal form if its rules follow the three following patterns: A $\rightarrow$B C, A $\rightarrow$ a, S $\rightarrow$ ϵ.

Where *A*, *B* and *C* are non-terminals other than the special symbol of the grammar *S*. *a* is a terminal and ϵ represents the empty symbol whose use is only allowed when the language generated by the grammar is considered as a well-formed string. The constraints imposed by this normal form are that all structures (syntax trees) associated with sentences generated by a grammar in Chomsky normal form are strictly binary in their non-terminal part.

If we go back to the grammar in Figure 4.20, we note that only the rule: B→ a B b violates the diagrams imposed by Chomsky's normal form, because its right-hand side contains more than two symbols. Thus, to make our grammar compatible with Chomsky's normal form, we make the changes provided in Figure 4.20.

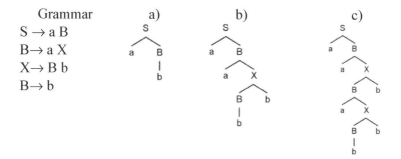

Figure 4.20. *Example of a grammar in Chomsky normal form with examples of syntax trees*

Note that in linguistic grammars, to avoid the three-branches rules as the one provided in Figure 4.21: NP → Det N PP, some assume the existence of a unit equivalent to a simple phrase that is called group: noun, verb, adjective, etc. (see Figure 4.21).

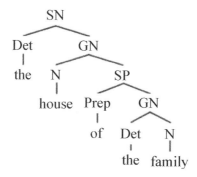

Figure 4.21. *Syntax tree of an NP in Chomsky normal form*

Grammar is said to be in Greibach normal form if all of its rules follow the two following patterns [GRE 65]: $A \rightarrow a$ and $A \rightarrow aB_1 B_2 \ldots B_n$,

Let us consider the grammar in Figure 4.22 for an example of the grammar in Greibach normal form.

$$
\begin{array}{l}
S \rightarrow aB \\
bA\ A \rightarrow a \mid aS \mid \\
bAA\ B \rightarrow b \mid bS \mid aBB
\end{array}
$$

Figure 4.22. *Example of grammar in Greibach normal form*

It should be noted that there are algorithms to convert any Type-2 grammar in a grammar in Chomsky or Greibach normal form. We should also note that beyond their theoretical interest, some parsing algorithms require standardized grammar.

Type-3 grammar, which is sometimes called regular grammar, is a type of grammar whose rules follow the two following patterns: $A \rightarrow a$ and $A \rightarrow a B$.

Where A and B are non-terminal symbols, whereas a is a terminal symbol. This is a typical G grammar of regular languages which generates the language: $L(G) = a^n b^m$ with n, m > 0 (see Figure 4.23).

$$
\begin{array}{l}
S \rightarrow a\ S \\
S \rightarrow a\ B \\
B \rightarrow b\ C \\
C \rightarrow b
\end{array}
$$

Figure 4.23. *Regular grammar that generates the language $a^n b^m$*

Finally, we should mention that it is formally proved that the languages generated by a regular grammar can also be generated by an finite-state automaton.

A particular form of regular grammar has been named "Type-4 grammar". The rules in this grammar follow the diagram: $A \rightarrow a$. In other words, no non-terminal symbol is allowed in this type of grammar. This grammar, which is of limited usefulness, is used to represent the lexicon of a given language.

Table 4.5 summarizes the properties of the types of grammar that we have just reviewed.

Type	Form of rules	Typical example	Equivalent model
0	$\alpha \rightarrow \beta$ (no limits)	Any calculable function	Turing machine
1	$\alpha A \beta \rightarrow \alpha \gamma \beta$	$a^n b^n c^n$	Augmented Transition Networks (ATN)
2	$A \rightarrow \gamma$	$a^n b^n$	Recursive Transition Networks (RTN)
3	$A \rightarrow a$ $A \rightarrow aB$	$a* = a^n$	Finite-state automata

Table 4.5. *Summary of formal grammars*

The presentation of the different categories of formal languages and their principal characteristics leads us to the following fundamental question: what type of grammar is able to represent all the subtleties of natural languages? To answer this question, we are going to discuss the limits of each type based on Type-3 grammar.

From a syntactic point of view, natural languages are not regular because some complex sentences have a self-embedded structure which requires Type-2 grammar for their processing (examples [4.29]):

- The mouse likes the cheese.
- The mouse the cat chased likes the cheese.
- The mouse the cat the rat bit chased likes the cheese.   [4.29]
- The mouse the cat the rat the lion looks at bits chases likes the cheese.

To understand the differences between the abstract structures of complex sentences, let us look at Figure 4.24.

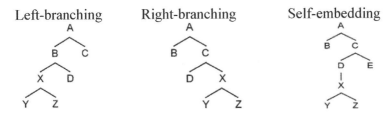

Figure 4.24. *Types of branching in complex sentences*

The syntax of natural languages is not independent of context because, in a language such as French, there is a multitude of phenomena, whose realization requires the consideration of the context. For example, we can mention the following: agreement, unbounded dependencies and passive voice.

Agreement exists between certain words or phrases in person, number and gender. In a noun phrase, the noun, the adjective and the determiner agree in gender and number. In a verb phrase, the subject and the verb agree in person and number. To integrate the constraints of agreement in a Type-2 grammar, we must diversify the non-terminals and then create as many rules for the possible combinations. The rule S → NP VP will be declined in six different rules as in the context-free grammar shown in Figure 4.25.

S → NP_Pers1_NoSing	NV_Pers1_NoSing
S → NP_Pers2_NoSing	NV_Pers2_NoSing
S → NP_Pers3_NoSing	NV_Pers3_NoSing
S → NP_Pers1_NoPlur	NV_Pers1_NoPlur
S → NP_Pers2_NoPlur	NV_Pers2_NoPlur
S → NP_Pers3_NoPlur	NV_Pers3_NoPlur
NP_Pers1_NoSing → I	
NP_Pers2_NoSing → you	
NP_Pers3_NoSing → he/she/it	
etc.	

Figure 4.25. *Type-2 grammar modified to account for the agreement*

This solution significantly reduces the generative power of grammar and makes its management very difficult in practice, because non-terminals are difficult to be read by humans. Some cases of agreement are even more complicated, as the agreement between the post-verbal attributive adjective of an infinitive and the subject noun phrase [4.30]:

Jane wants to become a top linguist.   [4.30]

Type-2 grammar does not allow us to specify the sub-categorization of a verb predicate (the required complements) or to require that the verb does not have complements. Here a possible solution is also to vary categories. These types of grammar do not allow the expression of structural generalizations such as the relationship between the passive and the active voice.

4.2.4. *Feature structures and unification*

The formalisms that we have just seen allow us to represent the information on the hierarchical dependencies of the constituents and their order. We have seen that with the grammar in Figure 4.25, it is quite important to take into account the relationships of agreement between these different constituents with a simple phrase structure grammar. The most current solution in the field of linguistics to solve this problem is to enrich the linguistic units with feature structures (FS), which contain information of different natures (morphological, syntactic and semantic) and to express the possible correlations among them. They also allow us to refine the constraints at the lexicon level and, thus, to simplify the rules of grammar. For more information on feature structures and unification, we refer to [SHI 86, CHO 91, WEH 97, SAG 03, FRA 12].

As we have seen in the sphere of speech, the first use of features in modern linguistics was proposed in the field of phonology, particularly with studies by Roman Jacobson, Noam Chomsky and Morris Halle. Since the 1980s, FSs have been adopted in the field of syntax, particularly in the framework of formalisms based on unification. It is a family of linguistic formalisms that allows us to specify lexical and syntactic information. The main innovation of these formalisms lies on two points: the use of features for encoding of information and the unification operation for the construction of higher constituents than the word level.

The unification operation is applied on features in order to test, compare or combine the information they contain. The origin of the unification concept is double: on the one hand, it comes from the studies on the logic programming language "Prolog" [COL 78, CLO 81] and on the other hand, it is the result of studies in theoretical and computational linguistics on the *functional unification grammar* (FUG) [KAY 83], the *lexical functional grammar* (LFG) [BRE 82, KAP 83] and the *generalized phrase structure grammar* (GPSG) [GAZ 85].

We distinguish two types of feature structures: atomic feature structures in which all features have a simple value and complex feature structures (CFS) where features can have other feature structures as a value. Below is the structural feature of the noun "house" and of the verb "*love*" which are shown in Figure 4.26. For example, the verb (FS b) has two features: an atomic feature *VerbType* and a complex feature *Agreement*.

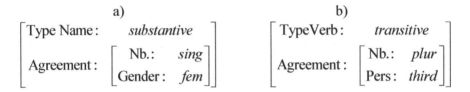

Figure 4.26. *Feature structures of the noun "house" and of the verb "love"*

Beyond the words, all other levels, such as phrases, clauses and sentences, can be enriched by CFSs. To clarify the enrichment of supra-lexical units with features, let us take a simple sentence such as [4.31]:

<u>Frank</u> eats <u>apples</u>.   [4.31]

As there is no consensus as to the necessary features for the analysis of a sentence, the linguistic theories which adopt the features as a mode of expression of linguistic properties differ on this point. Thus, we can adopt the following features to analyze the sentence [4.31]:

– *category:* specifies the grammatical category of the sentence: S;

– *head:* corresponds to the head of the sentence which is the verb and specifies the features: tense, voice, and number;

– *subject:* noun phrase which is an argument of the verb;

– *object:* noun phrase which is an argument of the verb.

This provides the CFS presented in Figure 4.27.

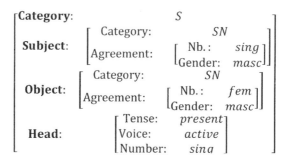

$$
\begin{bmatrix}
\textbf{Category:} & S \\
\textbf{Subject:} & \begin{bmatrix} \text{Category:} & SN \\ \text{Agreement:} & \begin{bmatrix} \text{Nb.:} & sing \\ \text{Gender:} & masc \end{bmatrix} \end{bmatrix} \\
\textbf{Object:} & \begin{bmatrix} \text{Category:} & SN \\ \text{Agreement:} & \begin{bmatrix} \text{Nb.:} & fem \\ \text{Gender:} & masc \end{bmatrix} \end{bmatrix} \\
\textbf{Head:} & \begin{bmatrix} \text{Tense:} & present \\ \text{Voice:} & active \\ \text{Number:} & sing \end{bmatrix}
\end{bmatrix}
$$

Figure 4.27. *CFS of a simple sentence*

A variant of the feature structures was proposed by Aït-Kaci [AIT 84] where each structural feature has a type which limits the features that can be included, as well as the values that the atomic features can have. This concept is comparable to the types in the framework of the object-oriented programming. For example, the complex feature *subject* can understand the category and agreement features and the atomic feature *Nb.* can have the values: *sing* and *plur*.

We should also note that CFSs can take the form of feature graphs, such as the examples of the feature structures presented in Figure 4.28.

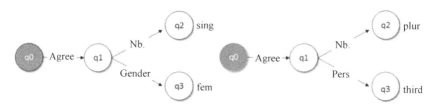

Figure 4.28. *Feature graphs for the agreement feature for the words "house" and "love"*

As we see in Figure 4.28, feature graphs are oriented, as we cannot navigate in the direction of the arrow (or arc). As these graphs are a data structure whose mathematical properties are well-known, this allows us to improve our understanding of CFS properties. Such a graph can be seen as a quadruplet: $\{Q, q', \delta, \theta\}$. Let us look at the graph (a) in Figure 4.28 as an example, to make our explanation more concrete.

– Q is a finite set of nodes: $Q = \{q_0, q_1, q_2, q_3\}$;

– $q' \in Q$ is the initial state of the graph $q' = q_0$. As we can note, in Figure 4.28, the initial state of this graph is colored to differentiate it from other states;

– δ represents a partial function in the graph such as: $\delta(q_0, \text{Agreement})= q_1, \delta(q_1, \text{Nb.})= q_2, \delta(q_1, \text{Gender})= q_3;$

– θ represents a leaf that corresponds to a feature. Thus: $\theta(q_2)=\text{sing}, \theta(q_3)=\text{fem}.$

The representation of the features in the form of a graph leads us to the path concept which is intimately associated with this. A path is a sequence of features used to specify a particular component of a structural feature. The paths that interest us are those that connect the initial element and a leaf. If we accept that a path is part of our CFS: $\pi \in \text{CFS}$, we can say that our function $\delta(q, \pi)$ provides the value of path π from the node q. Now, if two different paths begin with the initial state of the graph and ultimately lead to the same node, we then say that there is a reentrancy relationship between these two paths. Formally, we can express this as: $\delta(q_0, \pi) = \delta(q_0, \pi')$ and $\pi \neq \pi'$. By extension, a structural feature is called reentrant if it contains two features that share a common value. It is appropriate to distinguish the reentrancy of cases where two different paths lead to two different features which occasionally have the same value. In this kind of case, we refer to paths of similar values. Let us consider the examples of CFSs in Figure 4.29 to clarify this distinction.

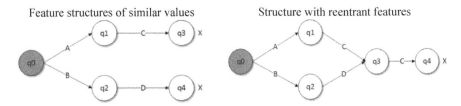

Figure 4.29. *Example of structures of shared value and of a reentrant structure*

As we can see in Figure 4.29, structures of similar values consist of an occasional resemblance between the values of different features (q_3 and q_4). This means that it is not a constraint imposed by grammar but a simple coincidence. A current example is when two noun phrases, subject and object direct, share the same features of agreement in gender and number. By contrast, the reentrant features are the shared product of two different paths of the same feature (q_4) as in the constraints on the agreement between two different constituents as the subject noun phrase and the verb phrase.

In matrix form, reentrant features are marked with specific indices. Thus, the symbol $\boxed{1}$ in Figure 4.30 means that the features f and g must have the same value.

Feature structures of similar values

$$\begin{bmatrix} f: & [h:a] \\ g: & [h:a] \end{bmatrix}$$

Structure with reentrant features

$$\begin{bmatrix} f: & \boxed{1}[h:a] \\ g: & \boxed{1} \end{bmatrix}$$

Figure 4.30. *Example of structures of shared value and of a reentrant structure*

If we approach the issue of the comparison of two feature structures, a key question arises: are there any CFSs which are more generic than others? To answer this question, the concepts of subsumption and extension have been proposed.

Subsumption is a partial order relationship that we can define on feature structures and which focuses both on the compatibility and the relative specificity of the information they contain. It is noted with the help of the symbol: $\sqsubseteq$. Let us look at the two feature structures FS and FS'. We say that FS is a subsumption of FS' if all information contained in FS is also present in FS'.

In Figure 4.31, we have three CFSs that maintain the following subsumption relationships: $a \sqsubseteq b$ and $a \sqsubseteq c$ and $b \sqsubseteq c$. In other words, as the CFS a is the least specific and consequently more abstract, it subsumes the structures b and c which are more generic. Obviously, the subsumption between these structures would not exist without the compatibility between these three structures. If we go back to Figure 4.26, the feature of agreement of the structure (a) does not subsume the feature of agreement of (b) and vice versa, because of the incompatibility between the two features: the feature *gender* in (a) and the feature *pers* in (b).

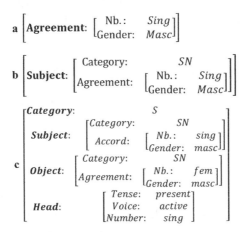

Figure 4.31. *Examples of feature structures with subsumption relationships*

Extension is the inverse relationship of subsumption. It can be defined in the following manner: let us consider two structures FS and FS'. We can say that FS' is an extension of FS, if and only if: all atomic features of FS' are present in FS with the same values, and if for all non-atomic features t_i, in FS', there is an atomic feature t_i' in FS such that the value of t_i' is an extension of the value of the feature t_i.

Naturally, all feature structures are not in a relationship of extension or subsumption. In fact, some CFSs can contain different but compatible information, such as the relationship between the object and the subject in matrix (c) as shown in the Figure 4.25. In addition, the structures "subject" and "head" are both different and incompatible.

We have seen that the syntactic analysis process consists of combining the representations of syntactic units to achieve a representation of the structure of the sentence. We have also seen examples of syntactic operations which allow us to combine words and phrases, etc. The question which arises at this stage is: how can we combine the representations of constituents enriched with features? The answer to this question is to use the unification operation. It is an operation which determines, from a set of compatible structures, a structure that contains all the information present in each of the members of the set and nothing else. The unification of the two feature structures FS_1 and FS_2 has, as a result, the smallest structure FS_3 which is an extension both of FS and FS'. If such a structure does not exist, then the unification is indefinite. Symbolized by the union operator U, unification can also be reformulated:

FS_1 U $FS_2 = FS_3$ where $FS_1 \subseteq FS_3 \wedge FS_2 \subseteq FS_3$.

Note that the unification operation is both commutative and associative:

FS_1 U $FS_2 = FS_2$ U FS_1

$(FS_1$ U $FS_2)$ U $FS_3 = FS_1$ U $(FS_2$ U $FS_3)$

As we can see in Figure 4.32, unification can be performed in quite varied configurations. In case (a), it is clear that the unification of a structural feature with itself is possible. Case (b) shows how we can unify the two different structures to obtain a richer structure. Case (c) shows that logically the unification of two incompatible feature structures is impossible because it leads to an inconsistency. In case (d), we see that the null structure acts as a neutral element and unifies with all structures without modifying them. Case (e) shows how we can perform unification with reentrant features. Finally, in case (f), we see a unification operation of more complex linguistic structures.

It should be noted that there are several generic tools that provide implementations of the unification process and which are available to the community. Among the best-known we can mention: PATRII, Prolog, and NLTK.

The logic programming language "Prolog", as we have seen, is the tool which inspired the first studies and is still valid. In fact, the unification of terms which are provided natively in Prolog significantly reduces the application development time.

PATRII is another tool available to the community in the field of unification. Originally proposed in SRI International by Stuart Shieber, PATRII is both a formalism and a programming environment written in Prolog [SHI 87]. It is possible to use PATRII to implement a limited variety of formalisms. It is based on Type-2 grammars with which FS are associated.

a. $[Nb.: sing] \cup [Nb.: sing] = [Nb.: sing]$

b. $[Nb.: sing] \cup [Gender: masc] = \begin{bmatrix} Nb.: & sing \\ Gender: & masc \end{bmatrix}$

c. $[Nb.: sing] \cup [Nb.: plur] = faux$

d. $[\ ] \cup [Nb.: plur] = [Nb.: plur]$

e. $\begin{bmatrix} Agreement: & \boxed{1}[nb.: & sing] \\ Subject: & [Agreement: & \boxed{1}] \end{bmatrix}$

$\cup \ [Subject: \ agreement \ [Person: \ 3rd]]$

$= \begin{bmatrix} Agreement: & [1]\begin{bmatrix} nb.: & sing \\ Person: & 3rd \end{bmatrix} \\ Subject: & [Agreement: & \boxed{1}] \end{bmatrix}$

f. $\begin{bmatrix} \textbf{Head}: & \begin{bmatrix} Tense: & present \\ Voice: & active \\ Number: & sing \end{bmatrix} \\ \textbf{Subject}: & \begin{bmatrix} Category: & SN \\ Agreement: & \begin{bmatrix} Nb.: & sing \\ Gender: & masc \end{bmatrix} \end{bmatrix} \end{bmatrix}$

$\cup \ \begin{bmatrix} \textbf{Agreement}: & \begin{bmatrix} Nb.: & sing \\ cas: & nominative \end{bmatrix} \end{bmatrix} =$

$\begin{bmatrix} \textbf{Head}: & \begin{bmatrix} Tense: & present \\ Voice: & active \\ Number: & sing \end{bmatrix} \\ \textbf{Subject}: & \begin{bmatrix} Category: & SN \\ \textbf{Agreement}: & \begin{bmatrix} Nb.: & sing \\ Gender: & masc \\ Case: & nominative \end{bmatrix} \end{bmatrix} \end{bmatrix}$

Figure 4.32. *Examples of unifications*

4.2.5. *Definite clause grammar*

Definite clause grammar (DCG) is a logical representation of linguistic grammars. The first form of this grammar, which was called "metamorphosis grammar", was introduced in 1978 at the University of Marseille following Alain Colmerauer's studies, of which the first application was on automatic translation. Afterward, David Warren and Fernando Pereira of the University of Edinburgh proposed a particular case of metamorphosis grammars which were named DCG [PER 80]. DCGs were created to develop and test grammars on the computer, particularly with the logic programming language "Prolog" and more recently with the language

"Mercury". From a functional point of view, DCGs allow us to analyze and generate string lists. Let us look at the DCG in Figure 4.33.

s --> np, vp.	n--> [boy].
s --> **p**, conj, **p**.	n --> [girl].
np --> det, n.	n --> [television].
vp --> v, np.	n --> [radio].
vp --> v.	v --> [looks].
det --> [the].	v --> [listens].
det --> [a].	conj --> [and].

Figure 4.33. *DCG Grammar*

The first note that we can make in respect of the grammar in Figure 4.31 is that its format is very close to the format of the ordinary Type-2 grammars, apart from a few small details. For example, non-terminals do not begin with a capital letter. As the latter is the indication of a variable according to the syntax of Prolog, terminals are provided in square brackets which are used in Prolog to designate lists. We also note that DCG allows the writing of recursive rules (the p symbol exists in the left and the right-hand side of the rule).

It is also necessary to add that DCGs can be extended to enrich the structures with features (see grammar in Figure 4.34).

```
s --> np(subj), vp.
np() --> det, n.
np(X) --> pron(X).
vp --> v, np(obj).
vp --> v.
det --> [the].
det --> [the].
n --> [father].
n --> [girl].
pron(suj) --> [he].
pron(suj) --> [she].
pron(obj) --> [him].
pron(obj) --> [her].
v --> [embraces].
```

Figure 4.34. *DCG enriched with FS*

Several extensions have been proposed to improve DCGs including XGs by Pereira [PER 81], the *definite clause translation grammars* (DCTGs) by [ABR 84] and the *multi-modal definite clause grammars* (MM-DCGs) by [SHI 95].

4.3. Syntactic formalisms

Given the interest in syntax, a considerable number of theories in this field have been introduced. Different reasons are behind this diversity, including the disagreement on the main units of analysis (morpheme, word or phrase), the necessary knowledge to describe these units, as well as the dependency relationships between them.

In this section, we retained three formalisms, including two that are based on unification: X-bar, HPSG and LTAG.

4.3.1. *X-bar*

Before introducing the X-bar theory, let us begin with a critical assessment of the phrase structure model presented in section 4.2.1. Let us look at the sentences [4.32] and the rewrite rule and the syntax tree of the sentence [4.32a] shown in Figure 4.35:

a) A pharmacist from Aleppo with the blue shirt.

b) A report from Istanbul on the Mediterranean.

c) A pharmacist with the blue shirt from Aleppo.*   [4.32]

d) A report on the Mediterranean from Istanbul.

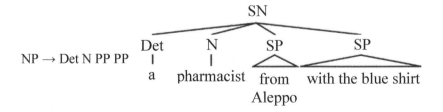

Figure 4.35. *Rewrite rule and syntax tree of a complex noun phrase*

The sentence [4.32a] mentions a pharmacist in the city of Aleppo who has a blue shirt and the sentence [4.32b] mentions a reportage of Istanbul about the Mediterranean. Intuitively, we have the impression that the two prepositional phrases in the two cases do not have the same importance. To demonstrate this, it is sufficient to perform a permutation between the two phrases in each of the sentences. This provides the sentences [4.32c] and [4.32d] which are not semantically equivalent, initial sentences ([4.32a] and [4.32b]) whose grammaticality is questionable. Thus, we can conclude that the two prepositional phrases do not have the same status. However, the plain phrase structure analysis proposed in Figure 4.35 assigns equal weight to them. Consequently, a revision of our model seems necessary in order to account for these syntactic subtleties.

Moreover, in the framework of a universal approach to the modeling of the language, we observe that the rules that we have proposed up to now are specific to the grammars of European languages such as English or French and do not necessarily apply to other languages such as Arabic, Pashto or Portuguese.

These considerations, among others, led to the proposal in 1970 of a modified model of the generative and transformational grammar which was named X-bar theory. Initiated by Noam Chomsky, this theory has been developed afterwards by Ray Jackendoff [CHO 70, JAC 77]. It allows us to impose restrictions on the class of possible grammatical categories while allowing a parallel of these latter elements thanks to metarules (generalization of several rules). It is also a strong hypothesis on the structure of constituents across languages. It rests on two strong hypotheses:

– all phrases, regardless of their categorical nature, have the same structure;

– this structure is the same for all languages, regardless of the word order.

The term X-bar is explained as follows. The letter X corresponds to a variable in the general diagram of the structure in constituents, which applies to all syntactic categories (N, V, A, S, Adv., etc.). The term bar refers to the notation adopted by this formalism to differentiate the fundamental levels in the analysis. They are noted with one or two bars above the categorical symbol of the head. Usually, for typographical reasons, we replace the notation bar with a prime notation. N", V", A" and S" are thus notational variants respectively of NP, VP, ITS, PP. The notation that we have adopted and the alternatives in the literature are presented in Table 4.6.

Level	Our notation	Alternatives
Phrase level	NP	$N", N", \overline{\overline{N}}, N^{max}$
Intermediate level	N'	$\overline{N}$
Head/Word level	N	N^0

Table 4.6. *Adopted notation and variants in the literature*

In the framework of the X-bar theory, a phrase is defined as the maximum projection of a head. The head of a phrase is a unique element of zero rank, word or morpheme, which is of the same category as the phrase as in the grammar of the Figure 4.36.

$$NP \rightarrow N\ PP$$
$$VP \rightarrow V\ NP$$
$$PP \rightarrow P\ NP$$
$$SA \rightarrow A\ PP$$

Figure 4.36. *Examples of phrases with their heads*

This leads us to a generalized and unique form for all rules: $SX \rightarrow X\ SY$, where X and Y are variables. Furthermore, according to the X-bar theory, a phrase accepts only three analysis levels:

– level 0 (X) = head;

– level 1 (X) = head + complement(s);

– level 2 (X) = specifier + [head + complement(s)].

The specifier is usually an element of zero rank, but sometimes a phrase can have this role: NP (spec(my beautiful) N^0 (flowers)). The specifier is a categorical property of the head word. For example, the determiner is a property of the category of nouns. Similarly, the complement is a lexical property of the head word: taking two object complements by the verb *give* is a lexical property of this specific verb, not a categorical property of the verb, in general. Let us examine the diagrams of the general structures of the phrases provided in Figure 4.37.

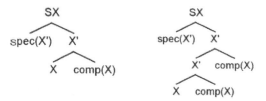

Figure 4.37. *Diagrams of the two basic rules*

In the two diagrams shown in Figure 4.37, we have the symbols X, Y and Z which represent the categories, Spec (X) which corresponds *to a specifier of* and Comp(X) for *complement of.* The categories *specifier* and *complement* designate the types of constituents. Thus, we have:

– Spec N = D (Det);

– Spec V = Vaux;

– Spec A = DEG (degree);

– Comp X' = N".

The hierarchical structure of the two syntactic relationships *specifier* and *complement* induces a constraint on the variation of order within the phrases. The X-bar theory predicts that the complement can be located to the left or to the right of the head and that the determiner is placed to the left of the "head+complement" group, or to the right of this group.

If we go back to our starting noun phrase *a pharmacist from Aleppo with the blue shirt* or even a variant of this phrase with an antepozed adjective to the noun phrase such as *good*, we obtain the analyses provided in Figure 4.38.

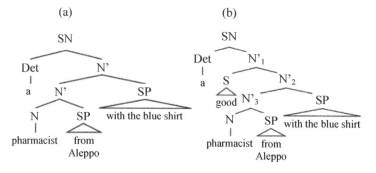

Figure 4.38. *Examples of noun phrases*

As we observe, the analyses proposed in Figure 4.38 take into account the functional difference between the two prepositional phrases. Similarly, the adjective responsible with respect to the head is processed as a specifier of the head of the phrase.

Although they appear to be acceptable, some researchers have doubted the adaptation of the analyses proposed in Figure 4.39 with respect to the principles of the X-bar theory. In fact, one of the principles, which form the basis of this theory, is to consider that all constituents, other than the head, must themselves be phrases. It is rather motivated by reasons related to the elegance of the theory than by linguistic principles. In the case of a noun phrase, this means that the only constituent that is not phrasal is the noun. However, in the proposed analysis, the determiner is analyzed similarly to the head. To resolve this problem, some have proposed an original approach which is to consider that NP depends in reality upon a determiner phrase whose head is the determiner [ABN 87]. The diagram of an NP according to this approach is provided in Figure 4.39.

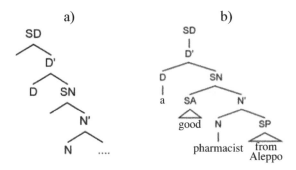

Figure 4.39. *Diagram and example of a determiner phrase according to [ABN 87]*

The processing of the verb phrase follows the standard diagram as can be seen in Figure 4.40.

After having shown how we analyze noun, verb and prepositional phrases in the framework of the X-bar theory, it is time to proceed with the analysis of an entire sentence. We have seen that it is generally easy to adapt the conventional analyses of constituents in the framework of the X-bar theory, but this is less obvious in terms of the sentence itself. Consider the conventional rewrite rule for the sentence: S → NP VP. The question that arises is to determine if the root of the sentence S is a projection of V (the head of VP) or a projection of N (the head of the subject NP). Two syntactic phenomena deserve to be examined prior to decide. Firstly, we must report the phenomenon of agreement between the subject NP and the VP in our sentence analysis. To do this, it would be useful to postulate the existence of a node between the subject NP and the VP, which would establish the link of agreement between these two phrases. Secondly, we can observe differences in the behavior of infinitive verbs and conjugated verbs. The two differ in the place they occupy with respect to the negation markers, certain adverbs and quantifiers. These differences confirm the special relationship which connects the conjugate verb with the subject noun phrase since it does not allow an interruption by these particular words. Invariably, the infinitive verb has no relationship of agreement with the subject (see [ROB 02] for some examples in French).

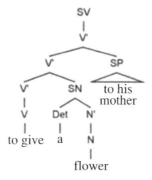

Figure 4.40. *Example of the processing of a verb phrase with the X-bar theory*

To account for the relationship between the subject NP and the VP, we can use an intermediate node between these two constituents. The role of this node is to convey the necessary information to the agreement of the verb AGG with the subject, as well as to provide the tense T. This node is called *Inflectional Phrase* (IP). We can now postulate that the sentence is a maximum projection of IP and that its head is, therefore, IP (see Figure 4.41).

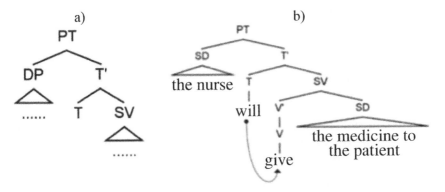

Figure 4.41. *Diagram and example of analysis of entire sentences*

The analysis of complex sentences is similar, in principle, to the analysis that we have already seen in the previous section. We assume the existence of a phrasal complement (PC) (see Figure 4.42).

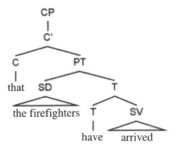

Figure 4.42. *Analysis of a completive subordinate*

Despite its elegance and its scientific interest, the X-bar theory has several limitations. Among them, we should mention the problem of infinitives and proper nouns, where it is difficult to identify the head.

The step that has followed the X-bar theory in the generative syntax was the proposal of the *government and binding* (GB) *theory* by Noam Chomsky at the beginning of the 1980s [CHO 81]. We refer to [POL 98, ROB 02,

CAR 06, DEN 13] for an introduction to this formalism, of which we will present only the general features.

To account for the different aspects of the language, GB is designed in a modular fashion involving a set of principles of which the most important are:

– government principle: this principle describes the phenomena of reaction. It addresses all conditions and constraints between the governors and those being governed;

– instantiation criteria of thematic roles: each lexical head associates the syntactic roles with their arguments (theta roles). Each argument of the sentence must receive a role and each and every role must be distributed. To receive a case, there must be a lexical or governed entity;

– binding principle: this principle describes the phenomena of co-referential anaphora.

In spite of its high complexity, GB formalism has been the subject of several operations in different application contexts [SHA 93, WEH 97, BOU 98]. GB formalism has, in turn, been the subject of amendments in the framework of Chomsky's minimalist program [CHO 95].

Finally, it is probably necessary to add that some concepts of the X-bar theory have also found their place in the theories related to generative grammar, including the HPSG formalism, to which we will devote the next section.

4.3.2. Head-driven phrase structure grammar

4.3.2.1. Fundamental principles

Head-driven phrase structure grammar (HPSG) was originally proposed to combine the different levels of linguistic knowledge: phonetic, syntactic and semantic. This formalism is presented as an alternative to the transformational model [POL 87, POL 96, POL 97]. Although it is considered as a generative approach to syntax, HPSG is inspired by several formalisms which come from several theoretical currents, including the *generalized phrase structure grammar* (GPSG), the *lexical functional grammar* (LFG) and the *categorial grammar* (CG). In addition, there are notable similarities between HPSG and *construction grammar* (CG) which can be noted. In fact, although CG focuses on essentially cognitive postulates, it has several points in common with HPSG, particularly with regard to the flexibility in the processing unit, as well as the representation of linguistic knowledge within these units [FIL 88] (see [GOL 03] and [YAN 03]). The legacy of these unification grammars, heirs themselves of several studies in artificial intelligence and cognitive sciences on the representation and the processing of knowledge, makes HPSG particularly suitable for IT implementations, which explains its popularity in the NLP community.

The processing architecture in the HPSG formalism is based on the single analysis level and does not therefore assume levels distributed as in the LFG formalism with the double structures: functional structure and structure of constituents. This is the same concerning the transformational model which assumes the existence of a deep structure and of a surface structure, or even as the GB model that attempts to explain the syntactic phenomena with the movement mechanism.

From a linguistic point of view, HPSG is composed of the following elements:

– a lexicon: which groups the basic words which are, in turn, complex objects;

– lexical rules: for derivative words;

– immediate dominance patterns: for structures of constituents;

– rules of linear precedence: which allow us to specify word order;

– a set of grammatical principles: which allow us to express generalizations about linguistic objects.

Since the complete presentation of a formalism as rich as the HPSG is difficult to achieve in a small section such as this, we refer the reader to the articles by [BLA 95, DES 03], as well as to Anne Abeillé's book [ABE 93] (in particular, the chapters on the GPSG and HPSG formalisms). We also refer to Ivan Sag's book and his collaborators [SAG 03] which constituted the primary source of information for this section.

4.3.2.2. Feature structures

FSs have a main role in the framework of HPSG formalism. More specifically, it refers to typed features, which means that to be well-formed, an FS must contain all the features required by its type and that all these features must be well instantiated. The types are organized in a generic hierarchy whose highest element is a *sign*. Words, phrases, sentences and speech are represented as signs as recommended in Saussurean tradition. Each sign has phonetic, syntactic and semantic characteristics which are grouped in an FS, whose general diagram is provided in Figure 4.43. The use of types in the development of the lexicon makes the latter more compact, thus facilitating its management and processing, and assuring its consistency. On the processing plan, types allow us to control the unification operation and consequently to avoid its failure.

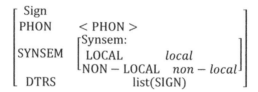

Figure 4.43. *Diagram of a typed FS in HPSG*

The *sign* feature has the function to indicate the status of the constituent at the head of which it is located. The *PHON* feature is used to list the words of the constituent. Exceptionally, it allows us to indicate the phonological properties of the constituent. The *SYNSEM* feature represents the syntactic and semantic features which describe the constituent. Sometimes, we prefer to have two separate features: SYN and SEM. The SYN feature provides the grammatical category of the node, its inflectional properties, as well as the nodes with which this element must be combined, whereas The SEM feature specifies the manner in which the sentence will be interpreted (mode of the sentence, participants, situation, etc.). *LOC* and *NON-LOC* features concern respectively the intrinsic/extrinsic aspects of the constituent which focus on its internal structure and its relationships with the objects that are located in its syntactic context. Most of the syntactic constructions are processed locally except for particular constructions, such as movement, which require non-local processing (see Table 4.7).

Type	Subtype
sign	word, phrase
phrase	headed phrase, headless phrase
list (σ)	nelist(σ) (nonempty list), elist (empty list) < >
set (σ)	neset(σ) (nonempty set), eset (empty set)
content	relationship, indexed-obj
relationship	gives-rel, walks-rel,...
Head	nomin, verb, adj, prep, ...
Case	noun, accus
Index	Per, Noun, Gender

Table 4.7. *Types in HPSG formalism [POL 97]*

Finally, DTRS node concerns child nodes and leads to two features: NON-HEAD-DTRS and HEAD-DTRS. To clarify the diagram provided in Figure 4.43, let us take the example of the FS in Figure 4.44.

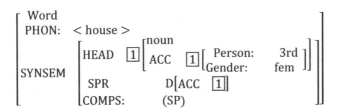

Figure 4.44. *Simplified lexical entry of "house"*

According to the FS provided in Figure 4.35, the word *house* is of the noun category and it has the features of agreement: 3rd person and singular.

It also says that this word must have a specifier (determiner), which has exactly the same features of agreement as the head, due to the use of a reentrant feature and an optional complement, placed between two parentheses, of a prepositional phrase type. Note that, in view of the wealth of FS in HPSG, many abbreviated forms or notational equivalents exist in the literature (see Figure 4.45 for a presentation).

Extended form	Abbreviated form
$\begin{bmatrix} \text{Head}: & \boxed{1} \\ \text{Tail}: & \boxed{2} \end{bmatrix}$	$\left\langle \boxed{1}, \boxed{2} \right\rangle$
$< \ldots\ldots \boxed{1} \, \| < >>$	$< \ldots\ldots \boxed{1} >$
$\begin{bmatrix} \text{Head}: & \boxed{1} \\ \text{Tail}: & \begin{bmatrix} \text{Head}: & \boxed{2} \\ \text{Tail}: & \boxed{3} \end{bmatrix} \end{bmatrix}$	$< \boxed{1}, \boxed{2} \mid \boxed{3} >$

Figure 4.45. *Some abbreviations of FS in HPSG*

The *S-ARG* or *ARG-S* feature concerns the argument structure which focuses on the relationships of binding of argument predicates. For example, verbs such as *see*, *love* and *write* can have the following argument structure: <see, [S-ARG < NP, NP>]>.

The FS of our lexical entry can be further enriched particularly with the addition of semantic features. Among them, we should mention MODE, INDEX and RESTR features. The MODE feature provides the semantic modes of an object and can have the values: question, proposal, reference, direction or none. The INDEX feature can take a theoretically unlimited number of values which correspond to the entities or the situation to which a constituent (or an entire sentence) can refer. Finally, the RESTR feature, whose value is a list of conditions, concerns conditions on the entities or the situation described by the INDEX feature, as well as the relationships that there may exist between them. If we take the word *love*, we have the following restrictions: the noun of the RELN relationship is *love* and this relationship implies two actors or entities: loving and loved; and in a specific situation, SIT. Similarly, the verb *give* implies a donor, a receptor and an object/gift always in a specific situation *s*.

Let us examine the two FSs of word inputs: *house* and *John* provided in Figure 4.46. In fact, this figure provides an enriched FS with S-ARG and SEM structure features of arguments. The S-ARG feature indicates that the specifier must be concatenated by the symbol $\oplus$ with the complementer. Semantic features indicate that this noun is of the type *reference* and that it has an index *k*. The proper noun *John*, meanwhile, has a simpler FS: the values of syntactic features, specifiers, complements and modifiers are empty. On the semantics level, it has a relationship of a proper noun (NP), which applies to a named entity.

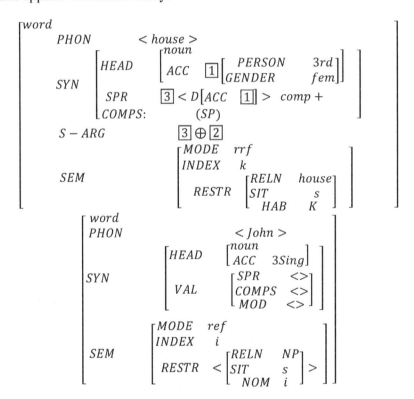

Figure 4.46. *Enriched FS of the words "house" and "John"*

With regard to verbs, features vary depending on the relationship of the verb with its arguments (transitive or intransitive verbs, etc.). Let us take the simplified examples shown in Figure 4.47.

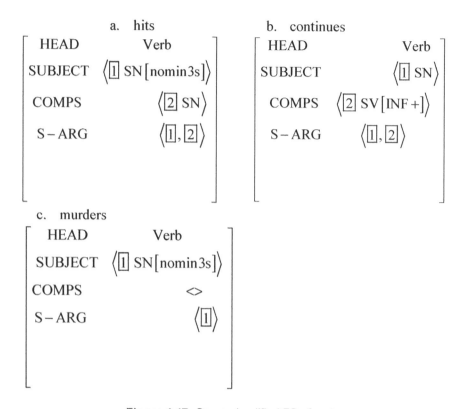

Figure 4.47. *Some simplified FS of verbs*

The transitive verb *hits* takes two arguments: a subject noun phrase and an object noun phrase at the nominative case and the third person singular. The verb *continues* takes a verb phrase as a complement. Finally, the verb dies, being intransitive, has an empty list of complement.

A more complete example of a verb is provided in Figure 4.48.

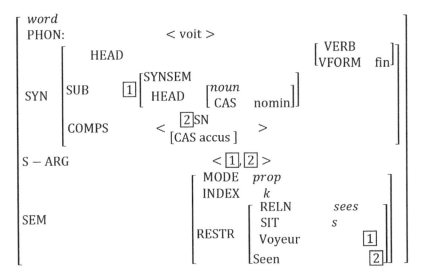

Figure 4.48. *FS of the verb "sees"*

The verb *sees* has two arguments: a subject noun phrase to the nominative case, which has the semantic role of *voyeur*, and an object noun phrase in the accusative case whose semantic role is *seen*.

4.3.2.3. *Morphological rules*

Lexical rules concern the introduction of morphological rules to FS of words in order to avoid redundancies, e.g. repeating the same FSs once for the singular form and once more for the plural form is expensive both in terms of grammar writing and in terms of memory, without mentioning the elegance of grammar and cognitive issues.

A lexical rule is an FS which establishes a link between two lexical sequences. We distinguish between two types of lexical rules *l-rules*: the rules of inflection *i-rules*, and the rules of derivation *d-rules*. All lexical rules obey the constraints described by the diagram in Figure 4.49.

$$\begin{bmatrix} \text{Input} & 1-\text{sequence} < \text{X}, [\text{SEM}/\boxed{2}] > \\ \text{Output} & 1-\text{sequence} < \text{Y}, [\text{SEM}/\boxed{2}] > \end{bmatrix}$$

Figure 4.49. *General diagram of l-rules*

Let us begin with i-rules. As we have seen in the sphere of morphology, inflection focuses primarily on the plural and singular forms for nouns, as well as the conjugation for verbs. To obtain a plural form for a noun from the singular form, we simply have to find the corresponding form according to the morphophonological rules that we have already seen, as well as to update the structural feature (see the example in Figure 4.50).

$$
\begin{bmatrix}
\text{INPUT} & < \boxed{1}, \begin{bmatrix} \text{noun} \\ \text{ARG} - \text{ST} & < [\text{compt+}] > \end{bmatrix} > \\
\text{OUTPUT} & < \text{Fnpl}\,\boxed{1}, \begin{bmatrix} \text{word} \\ \text{SYN} & \text{HEAD}[\text{ACC}[\text{NB Plur}]] \end{bmatrix} >
\end{bmatrix}
$$

Figure 4.50. *Rule of plural*

According to the rule illustrated in Figure 4.50, the plural form of a countable noun is a structure that keeps all the features of the singular structure, but with the NB feature of the agreement, which become plural. Obtaining the plural form of the noun is performed with the function of the plural noun F_{NPL}. The other forms of inflection and derivation follow the same principle. Let us take the case of the derivation of an agent noun from the verb: dance $\rightarrow$ dancer (Figure 4.51).

$$
\begin{bmatrix}
\textit{INPUT} & < \boxed{2}, \begin{bmatrix} Stv - lxm \\ SEM & [INDEX\ s] \\ ARG - ST & < X, & SN > \end{bmatrix} > \\
\textit{OUTPUT} & < F - eur\,\boxed{2}, \begin{bmatrix} cntn - lxm \\ SEM & [INDEX\ i] \\ ARG - ST & < Y \left(\begin{matrix} SPj \\ , [FORM\ de] \end{matrix} \right) > \end{bmatrix} >
\end{bmatrix}
$$

Figure 4.51. *Rule of derivation of an agent noun from the verb*

The function F-er adds an appropriate suffix to the form of the word at the output. The input of the rule is a verbal lexeme, whose INDEX feature of semantics becomes *i* at the noun form, because the change involves important semantic modifications resulting from the derivation, including the value of the MODE feature. By contrast, the value of the RESTR feature remains unchanged in both forms, because the information provided in the verbal form is compatible with the noun form. The ARG-FS feature requires a complement because, in languages such as English and French, we cannot get an agent noun from an intransitive verb: *give → giver, play → player,* but not *die → dier*. Followed by a genitive construction in French, an agent noun is located in a preposition phrase before the preposition *de* (of) as in *donneur de sang* (blood donor) and *joueur de football* (football player).

It is probably necessary to point out that, in cases where both mechanisms of derivation and inflection are necessary, we must first apply the rule of derivation and then the rule of inflection.

4.3.2.4. *Syntactic rules*

Unlike formalisms such as LFG and GB, the construction of the phrase is performed by combining and satisfying the constraints expressed by words. Several rules of word combination to form larger syntactic constituents are possible. We have selected the three rules that we considered to be the most important and which correspond to the basic rules of the X-bar theory.

The Head-Complement Rule takes the head *H* and produces a phrase if this head is located in a sequence of arguments which are compatible with the requirements for its COMPS feature. The FS of the obtained phrase is identical to the features of the head, except for COMPS features that are saturated (realized) and therefore removed by convention. In other words, in HPSG, the head produces a phrase which is similar in a manner compatible with the X-bar principle: the phrase is a maximum projection of the head. In addition, the labels $\boxed{1}....\boxed{n}$ must be the same in the following phrases. This means that the elements in the right-hand side of the rule must be selected by COMPS feature (see Figure 4.52).

$$\begin{bmatrix} \text{PHRASE} \\ \text{COMPS} \quad <> \end{bmatrix} \rightarrow \text{T} \begin{bmatrix} \text{WORD} \\ \text{COMPS} \quad < \boxed{1},....,\boxed{n} > \end{bmatrix} \boxed{1}....\boxed{n}$$

Figure 4.52. *Head-Complement Rule*

If we apply the rule of Figure 4.52 to a transitive verb and a noun phrase, this provides the tree of the Figure 4.53 where we have retained only the relevant features.

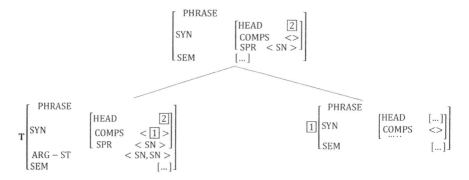

Figure 4.53. *Head-Complement Rule applied to a transitive verb*

We note in the tree structure of Figure 4.53 that the features of the head of the child node become the head features of the entire phrase. This is due to the Head Feature Principle (HFP) according to which the values of the head of any phrase are identical to the values of the head of the child node. We also note that the value of the SPR feature of the head of the child node is also transferred to the parent node. This is due to the Valence Principle (VP) which stipulates that the values of SPR and COMPS features must be identical to those of the head of the child node unless the rule specifies otherwise.

The Head-Modifier Rule allows us to obtain a phrase of a level equivalent to the X' in the X-bar theory. The MOD feature of the modifier must have the same value as the element it modifies (see Figure 4.54). Note that the HFP principle makes this rule generic (independent of the syntactic category), just as in the X-bar theory.

$$[\text{PHRASE}] \rightarrow \text{T}\boxed{1}[\text{COMPS} <>] \begin{bmatrix} \text{COMPS} & <> \\ \text{MODE} & <\boxed{1}> \end{bmatrix}$$

Figure 4.54. *Head-Modifier Rule*

The Head-Specifier Rule (Figure 4.55) takes a phrase with an SPR feature whose value is non-empty and combines it with an item that satisfies this value. The result is a phrase with a head preceded by a specifier.

$$\begin{bmatrix} \text{PHRASE} & \\ \text{SPR} & <> \end{bmatrix} \rightarrow \boxed{1}\text{T} \begin{bmatrix} \text{SPR} & <\boxed{1}> \\ \text{COMPS} & <> \end{bmatrix}$$

Figure 4.55. *Head-Specifier Rule*

4.3.2.5. *Semantic principles*

After having examined some syntactic rules for the combination of the constituents, we will now examine the semantic constraints which allow us to enrich syntactic structures with semantic properties. The two most important principles are: the Principle of Semantic Compositionality and the Semantic Inheritance Principle. The Principle of Semantic Compositionality stipulates that in any well-formed phrase, the RESTR feature of a parent node is the sum of the values of its child nodes. Similarly, the Semantic Inheritance Principle specifies that the values of MODE and INDEX features are identical to HEAD features of the child nodes. This means that in the HPSG formalism, semantics, just as syntax, is led by the head.

4.3.2.6. *Example of an analysis of a simple sentence*

To complete our presentation of the HPSG formalism, let us consider an example of the analysis of the following sentence: He sees the house. We will begin with the object noun phrase and to do this, we must first specify the FS of the determiner *the* as we have already seen the lexical entries of the noun *house* (Figure 4.46), as well as of the verb *sees* (Figure 4.48). Note that, for presentation conciseness reasons, we have seen fit to simplify the FSs.

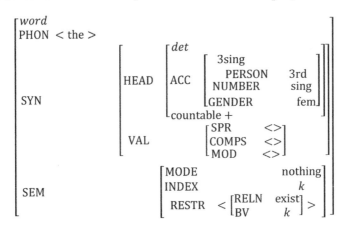

Figure 4.56. *Lexical entry of the determiner "the"*

The single element that deserves a comment in the FS of the Figure 4.56 is the Bound-variable (BV) feature. In fact, the semantic framework adopted for the quantifiers is the generalized quantifier theory which models both the standard quantifiers, such as the universal and existential quantifiers, and the non-standard quantifiers, such as *the most* or *the large majority* in the framework of the set theory.

As we can see in the tree of Figure 4.57, the FS of the determiner is identical to the FS of the determiner presented in Figure 4.56 with a few notable differences. On the one hand, the FS is marked by the symbol $\boxed{2}$, which allows us to insert the FS of the determiner *the* in the list of specifiers of the noun *house* allowing, according to the Head-Specifier Rule, the combination of these two words in the framework of the NP.

Similarly, the Semantic Inheritance Principle implies that the phrase inherits the MODE and INDEX features of the head node which are respectively: *ref* and *k*. This unifies the indexes of *the* and *house* and allows the determiner to quantify the noun in the framework of the noun phrase. Moreover, according to the principle of semantic compositionality, the value of a REST node of a parent node is the sum of the values of the child nodes: $\boxed{11}$, $\boxed{12}$.

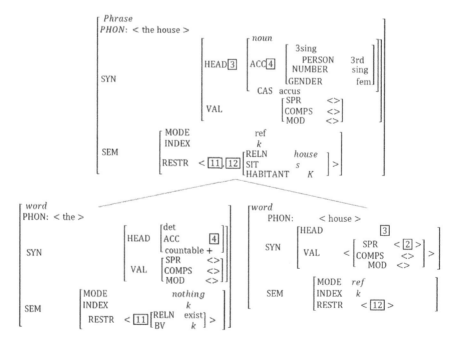

Figure 4.57. *Feature structures of the noun phrase: the house*

We are now ready to examine the analysis of the verb phrase: *sees the house* (Figure 4.58).

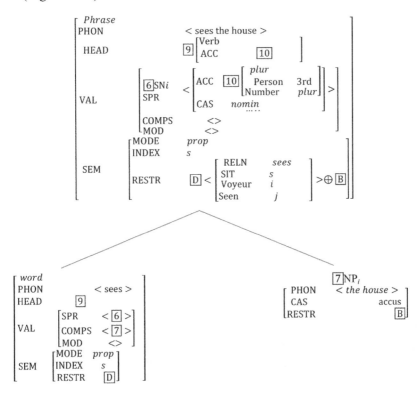

Figure 4.58. *Analysis of the verb phrase: sees the house*

Several observations can be made about the Figure 4.58. The noun phrase, labeled by the symbol $\boxed{7}$, whose FS is an abbreviation of the noun phrase of the Figure 4.57, is inserted in the list of the verb complements. Moreover, according to the Head-Complement Principle, the FS of the produced phrase is identical to that of the head of the complement, except for the COMPS feature whose value is empty (saturated). The value of the RESTR feature in the phrase is the concatenation of the values of the RESTR features in the child nodes: $\boxed{D}$ and $\boxed{B}$. According to the Valence Principle, the value of the SPR node of the root node is equal to the value of the SPR node of the head of the child node.

To analyze the entire sentence, we still have to introduce the FS of the subject pronoun, which is presented in Figure 4.59, as well as its combination with the structure of the Figure 4.60.

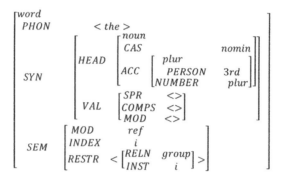

Figure 4.59. *The FS of the pronoun "the"*

To complete the analysis of the sentence, we still have to combine the structure constructed for the verb phrase (verb + complement NP) with the FS of the subject pronoun *he*, which provides the representation of Figure 4.60. The value of the RESTR node is the result of the concatenation of: $\boxed{E} \oplus \boxed{D} \oplus \boxed{B}$. The value of the AGR feature in the root node is equal to the value of the subject and in accordance with the Head Feature Principle (HFP) the Head-Specifier Rule, as well as the constraint of agreement between the verb and its specifier.

HPSG has been applied to several languages of various families, such as Arabic, French, German, Danish, etc. The *deep linguistic processing with HPSG* (DELPH-IN[3]) initiative has allowed the development of grammars of large sizes for languages such as English, German and Japanese. These grammars, which are available free of charge, are compatible with syntactic parsers such as LKB.

Note that HPSG has been used in real-time applications such as automatic speech translation, particularly in the German project Verbmobil [USZ 00, KIE 00]. Besides, generic software for HPSG grammars writing is also available, such as the LINGO system [BEN 02][4].

---

3 www.delph-in.net/.

4 www.lingo.stanford.edu/.

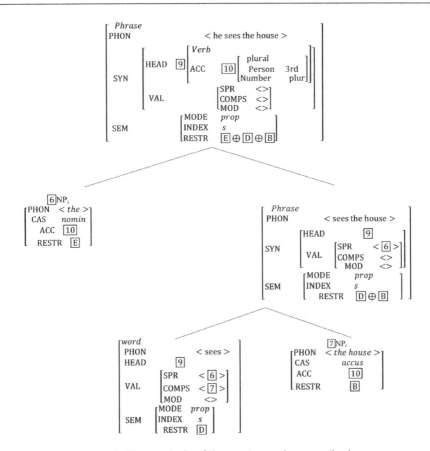

Figure 4.60. *The analysis of the sentence: he sees the house*

4.3.3. *Lexicalized tree-adjoining grammar*

4.3.3.1. *Fundamental principles*

The formalism of lexicalized tree-adjoining grammars has been described first in [JOS 75], under the initial name of *tree adjunct grammar*. It was then developed by other researchers particularly at the universities of Pennsylvania and Delaware in the United States, as well as at the University of Paris 7 in France (see [ABE 93] for a review of the development steps of this formalism). It is a lexicalized formalism which can be seen as an intermediate vision between the dependency grammars and the phrase structure grammars. In fact, this formalism is based on a representation of dependency relationships in the form of trees.

From a formal point of view, the LTAG formalism can be defined as a quintuplet (Σ, NT, I, A and S), where [JOS 99]:

– Σ is a finite set of terminal symbols;

– NT is a finite set of non-terminal symbols: $\Sigma \cap NT = \phi$;

– S is the distinguished non-terminal symbol: $S \in NT$;

– I is a finite set of trees called initial trees which are characterized by the following points: the internal nodes are labeled with non-terminal symbols, and the border nodes of initial trees are labeled by terminal and non-terminal symbols;

– A is a finite set of trees called auxiliary trees that have two fundamental properties: the internal nodes are labeled with non-terminal symbols and the nodes on the borders of auxiliary trees are labeled with non-terminal symbols.

From a functional point of view, LTAG can be described according to three points: the processing units (the elementary trees), the composition operations, as well as the features and their unification.

4.3.3.2. Basic units

Unlike conventional syntactic formalisms whose phrase constitutes the fundamental unit, LTAG has adopted a richer unit for the representation that is the elementary tree. Therefore, an LTAG grammar can be considered as a finite set of elementary trees. Any elementary tree has at least one of its leaf nodes occupied by a lexical item which acts as a head and that is generally called the anchor of this tree. The depth of the elementary trees is not limited to a branch[5]. In addition, two types of basic trees are distinguished in this formalism: the elementary trees and the auxiliary trees.

Elementary trees constitute a set of trees which are combined by substitution and which correspond to the basic syntactic structures. These trees are generally marked by the symbol, α.

Auxiliary trees are combined by adjunction. These trees have a leaf node, which is called foot node, bearing a non-terminal symbol in the same

---

5 The depth is the number of branches that separate the root node of the tree from the anchor of this tree.

category as the root node. Auxiliary trees are used for the representation of modifiers (adjectives, adverbs and relatives), completive verbs, modal verbs and auxiliary verbs. These trees are generally denoted by the symbol, β.

The leaf nodes of elementary trees can be annotated by terminal and non-terminal symbols. Two types of nodes annotated by non-terminal symbols can be distinguished: the substitution nodes marked by ($\downarrow$) and the adjunction nodes marked by (*). The construction of elementary trees conforms to four principles of well-formedness [ABE 93]:

1) Principle of lexical anchoring: each elementary tree must be associated with at least one lexical head. Unlike HPSG and other formalisms, the lexical head of an elementary tree in LTAG cannot be empty. In addition, an elementary tree can be anchored by a set of lexical items. In this case, we refer to co-heads. Co-heads are generally functional complementers, such as *from* and *that* (relative). Thus, each lexical entry is associated with all the structures that characterize its possible uses. From a computational point of view, lexicalization allows us to invoke only the subset of elementary trees of the grammar which is actually anchored by the words of the sentence, making the processing more effective.

2) Predicate-arguments co-occurrence principle: the syntactic relationships are compared to the logical relationships between the predicate and the argument. This means that any predicate must contain in its elementary structure at least a node for the arguments that it subcategorizes.

3) Principle of semantic consistency: any elementary tree must have a non-empty semantic representation.

4) Principle of non-compositionality: an elementary tree corresponds to a single semantic unit.

Semantic principles (3 and 4) are quite vague, since no clear definition is provided for what is meant by a semantic unit. Their role in LTAG is essentially to prevent most of the functional elements, prepositions, complementers, etc., to constitute principle autonomous elementary trees (2). Principle (3) serves to limit the size of elementary trees and to prevent the anchoring of some trees by unnecessary elements. A few examples of elementary trees in LTAG are provided in Figure 4.61.

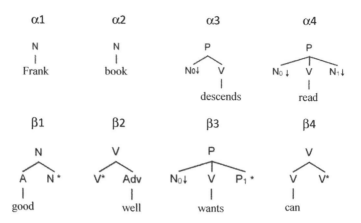

Figure 4.61. *Examples of initial and auxiliary elementary trees*

4.3.3.3. *Tree composition operations*

We can distinguish between two types of constraints on the composition of elementary trees within the LTAG formalism: syntactic constraints and semantic constraints. These different constraints influence the nature of the operations of the composition used. Two syntactic composition operations are possible: substitution and addition.

Substitution is similar to the rewrite operation in a Type-2 grammar. It allows us to insert a tree, initial or derived, to a substitution node of an elementary or derived tree which is noted by the sign: ↓. Substitution is an obligatory operation to a terminal substitution node. An example of substitution is the insertion of the initial tree of a determiner in the tree of a nominal group (Figure 4.62).

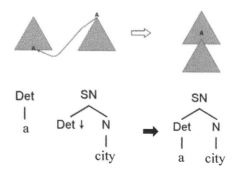

Figure 4.62. *Diagram and example of substitution in LTAG*

Addition is a specific operation in LTAG formalism. It allows us to insert an auxiliary (or a derivative of an auxiliary) tree to an internal node or a root node of an elementary or derived tree. Node *X*, where adjunction takes place, is replaced by an elementary tree whose root node and foot node are labeled by the category *X*. To illustrate adjunction, let us take as an example the insertion of the auxiliary tree which corresponds to the adverb at the internal node V of the initial tree of the verb *works* (Figure 4.63).

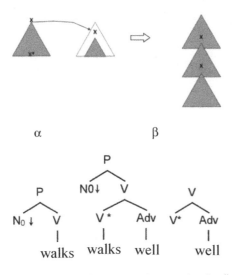

Figure 4.63. *General diagram and example of adjunction*

To control the adjunction in an LTAG grammar G = (Σ, NT, I, A, S), three types of constraints are defined on a given adjunction node [JOS 99]:

– Selective adjunction (SA (T)): this constraint allows the adjunction to single members of the set T $\subseteq$ A of auxiliaries trees at the given node. In this case, the adjunction of an auxiliary is not obligatory in this node.

– Null adjunction (NA): it prohibits any type of adjunction at the given node.

– Obligatory adjunction (OA(T)): this constraint requires any auxiliary tree which is a member of the set T $\subseteq$ A adjoined to the given node.

The composition process of basic units in larger units (or derivation) presents several specificities compared to other conventional syntactic formalisms. In fact, unlike phrase structure grammar of type CFG or another type, derivation is not characterized as a string obtained by other strings, but as a tree obtained from other trees. The direct result of this difference is the distinction within the LTAG formalism of two modes of representation of the result of the derivation, which are the derived tree and the derivation tree.

The derived tree is similar to the syntax tree in the phrase structure formalisms. It is a tree whose root is labeled with the distinguished symbol from the formalism and the leaves of which the lexical items of the analyzed utterance are aligned.

The derivation tree is a tree in which the nodes have pairs (elementary tree, address of the node of the higher tree where this tree has been inserted). The main function of the derivation trees is to show the dependencies between the lexical items. Note that in Type-2 grammars, the derivation tree and the derived tree are the same. An example of a derived tree and a corresponding derivation tree is provided in Figure 4.64.

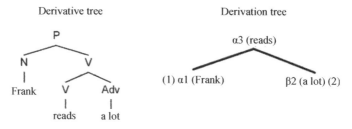

Figure 4.64. *An example of a derived tree and a corresponding derivation tree*

The derivation tree of the Figure 4.64 can be read as follows: the tree α1 (François) is substituted in tree α1(sleeps) at the location (1) address of the node N and tree β2 is adjoined to tree α1(sleep) at the node (2).

4.3.3.4. *Semantic composition and unification operation*

After having completed syntactic operations, we still have to integrate semantic constraints. To do this, LTAG proceeds to the decoration of the nodes of the syntactic trees with FSs. It refers to atomic structures that have the form (attribute, value). These features can be morphological, syntactic and semantic. They are defined at the level of elementary trees and must be retained in the derived trees. To facilitate the unification of features, LTAG distinguishes between two types of features that are present in each node: top features and bottom features. Top features show the relationships of the node with the nodes which dominate it, whereas bottom features are used to indicate the relationships of the node with the nodes that it dominates (see Figure 4.65).

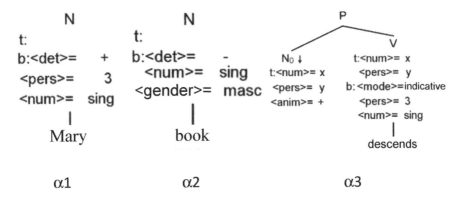

Figure 4.65. *Examples of feature structures associated with elementary trees*

In addition to the grouping of features, the unification operation enables us to express the constraints on the possible tree attachments. Thus, the two syntactic operations of the TAG formalism are constrained by the unification in two ways: unification with substitution and unification with adjunction.

In the case of substitution, top features of the root node of the substituted tree must be unified with the features of the node where there has been a substitution (see Figure 4.66).

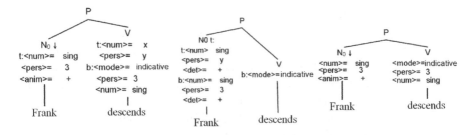

Figure 4.66. *An example of a substitution with unification*

In the case of adjunction, we must have, on the one hand, the unification of the top features of the root node of the auxiliary tree with the top features of the node receiving the adjunction, and on the other hand, the unification of the features of the foot node of the auxiliary tree with the foot features of the node receiving the adjunction (Figure 4.67).

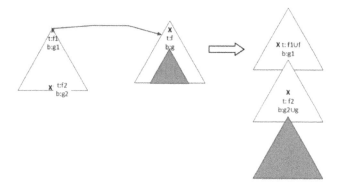

Figure 4.67. *Diagram of an addition with unification*

At the end of an analysis, for each complete derivation obtained, the top and bottom parts must be unified at each node of the corresponding derived tree.

Despite their usefulness for processing, the enrichment of the formalism with features is a fairly difficult task and requires a lot of work. Regarding the adaptation at the processing of oral dialogues, some of these features appear to be redundant and repetitive.

Having generated a lot of interest in the community, particularly thanks to its elegance and effectiveness, several researchers have proposed variants of this formalism. Sometimes these variants are motivated by the following reasons: to meet the needs of automatic translation – the parallel TAGs [SHI 90] and the parallel TFGs [CAV 98b], to simplify the parsing algorithm – Tree Insertion Grammar (TIG) [SCH 95], the stochastic TAGs [SCH 92, RES 92b], TFGs [DES 90, CAV 98a, ROU 99a], and to adapt to the processing of oral dialogues oriented by the semantic tree association grammar Sm-TAG [KUR 00].

Several parsing algorithms have also been proposed for LTAG of which we mention the parsing algorithm by the connectivity for the oral language [LOP 99], an efficient algorithm [EIS 00], a statistical algorithm [JOS 03] and a tabular algorithm [NAS 09]. An approach of shallow parsing based on LTAG has also been proposed under the name of supertagging [CHE 99, SRI 99]. Applications of this spoken language processing approach have been carried out [ROU 99b].

4.4. Automatic parsing

In the previous sections, we have seen several approaches for the syntactic description of a given language in the form of grammars. The question which now arises is how, from these grammars, to perform parsing of sentences. To answer this rather complicated question, several approaches have been proposed in the literature with a varied use of linguistic knowledge.

Parsing modules are of utmost usefulness in many fields of application, including automatic translation, grammatical correctors, man-machine dialogue systems, and sometimes speech recognition. Typically, parsing module provides its structural analysis to a semantic analysis module which must describe the semantic content.

The task of a syntactic parser is the construction of a syntactic representation, in the form of a parse tree or of another hierarchical data structure, from an input sentence and a grammar that describes the language. This task can be divided into three principal subtasks:

– *segmentation:* this step is to segment the string of words in phrases, sentences or any other morphological, syntactic chunks [ABN 91a, ABN 95], supertags, [SRI 99], semantic [KUR 00] or discursive unit [WEB 04].

– *categorization:* this step is to label the units obtained in the segmentation phase, a well-formed (syntactic or semantic) structure.

– *disambiguation:* in some cases, the parser can provide several parses corresponding to a single sentence. A disambiguation strategy is therefore required to obtain the best parsing tree. The main disambiguation methods include the use of linguistic metarules, psychological principles (minimal attachment, lexical preference, etc.) or probabilities [BOD 95].

Today, with the popularization of the use of NLP tools, including syntactic parsers in the processing of very large quantities of linguistic data, the complexity of parsing algorithms is taken seriously. In fact, some favor the use of approaches based on Finite State Automata (FSA) known for their effectiveness (complexity $O(n^2)$) rather than more traditional approaches such as the tabular algorithms whose complexity is equal to $O(n^3)$. With this importance, there are several parsers available to the community which are in reduced form as with the NLP toolkit (NLTK) [BIR 09], and the Stanford parser[6] which comes with several algorithms and several grammars for languages such as English [MAR 06], French [GRE 11], Arabic [PPE 10], Spanish and Chinese [LEV 03, CHA 09]. We can also mention the Charniak and Caballero [CAB 98] parser which is very popular for English, as well as the Xerox parser[7] which is available online for English, French and German.

Several good references are available in the literature on parsing algorithms. For example, the book by [AHO 88] focuses on the syntactic parsers used in the compilation of programming languages. [GRU 95] is a general book parsing techniques and formal languages whose first edition is freely available online[8]. [SIK 97] proposed a unified formal framework to describe and compare the different parsing algorithms. [CRO 96] is an introduction to the parsing presented in a cognitive perspective. Finally, [WEH 97] presents an introduction to parsing concepts both from a linguistic and an algorithmic point of view.

4.4.1. *Finite-state automata*

We have seen that finite-state automata (FSA) are a very good tool for modeling the phonological and morphological knowledge. We have also seen that regular grammars which are formally equivalent to the FSA are not rich enough to account for the subtleties of the syntax of natural languages including phenomena such as self-embedding. Yet, several studies have focused on the use of these tools for parsing in contexts where a partial modeling of syntactic knowledge seems to be sufficient, such as language modeling for speech recognition systems or robust parsing systems. The computational benefits behind the adoption of FSA are the relative

---

6 www.nlp.stanford.edu:8080/parser/.

7 www.open.xerox.com/Services/XIPParser/Consume/Parse%20text-64.

8 www.dickgrune.com/Books/PTAPG_1st_Edition/BookBody.pdf.

simplicity of implementation of these automata and the advantageous computational cost, as they are linear approaches. To obtain a finite-state automaton from approximation algorithms, several approaches have been proposed such as the approach which is based on the left-corner algorithm [JOH 98].

Several robust parsing approaches have been presented, of which the best-known is the chunk analysis approach by Steven Abney, which uses a cascade of FSAs [ABN 91a] (see section 4.4.10.1 for more details). A parsing approach of the oral language has also been proposed by [KAI 99].

4.4.2. *Recursive transition networks*

RTNs are an extended version of AFE [WOO 70]. Just as AFEs, they are composed of a series of states and transitions, and on a labeled graph in which each label corresponds to a category (lexical, syntactic or conceptual), the transition from one state to another is subordinated by the success of the unification between the label of the arc and the current word (or subnetwork). We could also refer to the philosophical and cognitive presentation of the recursion concept and of RTN in the book by [HOF 99].

Thus, a state in an RTN consists of four elements:

– *current node*: this element provides information on the processing location;

– *the rest of the sentence*: indicates the part of the sentence which has not yet been processed;

– *the nodes on hold*: the nodes in the current network which have not yet been crossed;

– *parse*: the parse associated with the processed part of the input sentence.

Three actions are possible when the parser is in a particular state according to the nature of this state:

– *the label is a phrase category (subnetwork)*: place the current node in the waiting stack and create a new constituent for a new category;

– *the label is a lexical category*: verify the identity of this word and add this word as well as its category to the current constituent;

– the constituent is complete: take the node on hold from the stack and integrate the current constituent in a higher level constituent.

In a more formal way, a string S composed of a set of substrings $s_1 \ldots s_k$, such as $S = s_1 \ldots s_k$, is recognized as C by a network N if and only if:

– C is the label of an arc that connects an initial state x and a final state y (where x and y correspond, respectively, to 1 and k);

– there is a path (a string of labels) $l_1 \ldots l_k$ accepted by N (seen as a non-RTN) and with x as the initial state;

– for each s_i (where $k \geq i \geq 1$), is $s_i = l_i$ (in this case s_i corresponds to a word) or s_i is recognized as a subnetwork l_i.

Thus, unlike phrase structure grammars, which consist of linear series of symbols, RTNs constitute a symbol lattice. To account for the components of the symbol lattice created by an RTN, we have adopted the notation of the Table 4.8 which is inspired by [GAV 00].

Notation	Arc type
λ_{SR}	The beginning of the sequence of a rule
$\lambda^{-1}{}_{SR}$	The end of the sequence of a rule
λ_{RA}	The beginning of the alternatives to a rule
$\lambda^{-1}{}_{RA}$	The end of the alternatives to a rule
λ_{TAV}	The empty forward transition
λ_{TArV}	The empty backward transition

Table 4.8. *Labels adopted for the annotation of RTN*

Here is an example of a transition network presented with the notation that we have adopted:

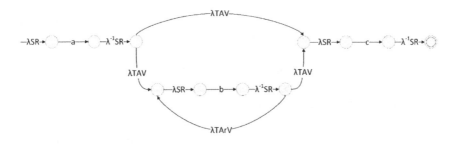

Figure 4.68. *Example of a recursive transition network*

This network allows us to recognize strings such as: *ac* (the empty backward transition allows us to not consider *b*), *abc*, *abbc* (the transition after an empty element allows us to accept an infinite number of *b*), *abbbc*, etc.

Although they are equivalent to the CFGs, RTNs have several advantages compared to them. In fact, RTNs are more compact and more effective than conventional phrase structure rules, since an RTN can cover several rules. To clarify this idea, let us look at the small grammar of Figure 4.69 in DCG format. For the sake of concision, we have omitted the rules in which the right-hand side is a terminal symbol.

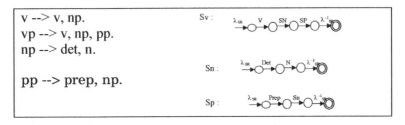

Figure 4.69. *A DCG and the corresponding RTNs TRVIDF PP*

The first note that we can make about DCG of the Figure 4.69 and the equivalent RTNs is that the rules corresponding to the NP are established within a single network. In addition to the advantage of the conciseness of this representation, processing with an RTN is more efficient than with the phrase structure grammar. Suppose that we want to analyze the phrase: *The dog chases the cat near the elephant*. With a top-down algorithm which uses the phrase structure grammar, firstly the system tries the first rule in which the left-hand side is v (VP-->v, np) and as the totality of the utterance is not yet parsed, it tries the second alternative of VP which also includes a *PP*. The problem is that with the second attempt, the system must restart the parsing of the verb and of the NP which were correctly analyzed the first time. As the two alternatives of the NV are coded with a single network, the two elements which are common to the two rules of the NV are retained when the system tries to verify the non-shared elements between these two forms. This property makes the RTNs comparable to tabular algorithms that we are going to see below. However, a notable difference between RTNs and tabular algorithms deserves to be mentioned. In fact, parse tables in tabular algorithms are created online (during the parsing), whereas in RTNs, the graph corresponding to the grammar is created offline during grammar compilation.

Another notable advantage of RTN is that it is easy enough to express the infinite repetition of any element in the grammar. This property is particularly useful for the implementation of a strategy of selective parsing, which is also called island (driven) parsing, where the system ignores the parts of the phrase that it is unable to parse.

RTNs have been a very popular paradigm in the 1970s and 1980s for parsing tasks. They have been used for the implementation of semantic grammars for the processing of oral dialogues (see among others [WAR 91, MAY 95, GAV 00]).

Finally, we note a form of transition networks which are even more powerful than RTN: Augmented Transition Networks (ATN) [VIN 83]. Two fundamental differences distinguish the ATR from the RTN. Firstly, ATRs have a data structure that is called a *register* which allows us to store the information and consequently to take the context into greater consideration. Secondly, ATRs allow us to define actions to perform at each arc transition. This allows ATRs to use them in order to implement the grammars based on unification.

4.4.3. *Top-down approach*

In this type of algorithm, the system constructs the parse tree by presupposing the existence of a sentence (or the axiom of the grammar). In other words, the system begins with the establishment of P at the root node of the parse tree. The second step is to find a rule with P in the left side and then to generate the corresponding branches at the categories on the right side of the rule as child branches of P. This procedure is repeated for the first branch until we arrive at a non-terminal symbol and the parser searches for another branch to explore.

Bottom-up parsers, like top-down parses, use a pointer to the next word to parse to keep a record of parsing progress. Both algorithms also use a stack. In Bottom-up algorithms the stack has a function of keeping track of the categories to find (hypotheses) while in top-down algorithms the stack is used to store the categories that are already included in the parse tree.

Let us consider the mini-grammar of the Figure 4.70.

PC → NP VP
NP → Det N
NP → NP
VP → V Adj
VP → VI
Det → the, my, my
NP → John
V → is | loves | descends
VI → descends
N → book | house | family
Adj → interesting | expensive

Figure 4.70. *Context-free grammars for the parsing of a fragment*

To analyze the sentence: *The book is interesting* with the grammar of Figure 4.70 using a top-down algorithm, we have the steps presented in Figure 4.71.

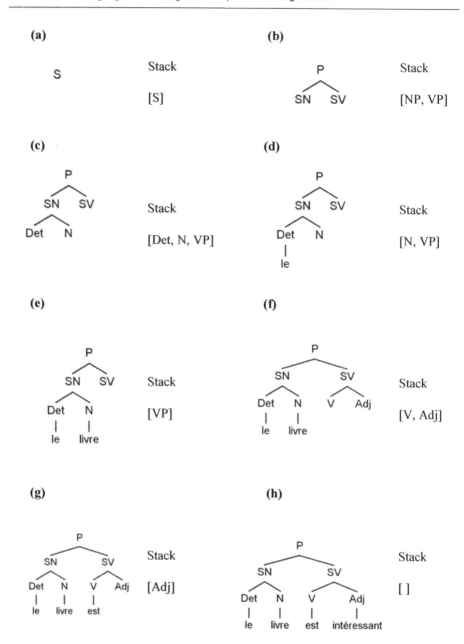

Figure 4.71. *Example of parsing with a top-down algorithm*

The top-down parsing algorithm is shown in Figure 4.72.

---

1. Initialization: stack = [S]
2. If the top of the stack (its first element) is a non-terminal N, then:
    *   Select a rule of the form N →β (where β is one or several symbols).
    *   Remove N from the stack
    *   Add β at the top of the stack.
3. If the element at the top of the stack is a preterminal T, then
    *   Find the next word of the sentence M
    *   If there is a rule of the form T → M, then remove T from the stack
    *   Otherwise, failure
4. If the stack is empty (stack =[]) and if there are no more words to analyze, then
    *   Success
5. Otherwise, go to step 2

Figure 4.72. *Basic top-down algorithms*

Step (1) of the algorithm initializes the stack with the special non-terminal symbol of the grammar, i.e. *S*. Step (2) searches for a rule in which the left-hand side is equal to the symbol which is located at the top of the stack. In the event of a success, it replaces this symbol by the non-terminal symbols that are located in the right-hand side of the rule. In our example (Figure 4.72), in step (b), the algorithm has searched for a rule in which the right-hand side is the symbol *S*, located at the top of the stack. It is the rule: S → NP VP. Then, *S* is removed from the stack and the symbols NP and VP are added. Step (3) deals with the case where a preterminal symbol, *T*, is located at the top of the stack. A preterminal symbol is a symbol which is rewritten in a terminal symbol. In other words, it is a morphological category (noun, determiner, verb, etc.). In this case, the algorithm tries to find a rule that has the preterminal symbol in its left-hand side and the next word of the sentence in its right-hand side. In the event of a success, the algorithm removes *T* from the stack; otherwise, it declares its failure. If we go back to our example in Figure 4.71, the preterminal symbol is Det and it is associated with success in the next word of the sentence *the* (in this case, it is the first) which is a determiner and this thanks to the rule (Det → the). Step (4) is the stop condition of the algorithm. In fact, for the algorithm to succeed, two conditions must be met: the stack must be empty and all the words in the input sentence must be parsed. In our example (Figure 4.71), the algorithm terminates with success in step (4) because, on the one hand, the stack is empty and, on the other hand, all the words in the sentence are parsed. The last step (5) allows the algorithm to loop until it arrives at a point where it declares its success or failure.

The question now is how our parsing algorithm will process syntactic ambiguities. In reality, both bottom and top parsing algorithms can adopt one of three exploration strategies to process ambiguities: backtracking, determinism and breadth-first search.

In our presentation of the top-down parsing approach, we have adopted an exploration strategy which is called *backtracking* or the *Depth-First Search*. This strategy is to develop a single rule even when the grammar offers several possible rules at a given point of the parsing. If the algorithm arrives at a parse that is not compatible with the sentence in the course of the parsing, it reverts to (it performs a *backtracking*) the first not explored alternative and it modifies the parse by adopting a new possible rule. To illustrate this mechanism, suppose that the input of our parsing algorithm is: John likes the book. The algorithm will first try to analyze the subject noun phrase with the rule (NP → Det N). Such an attempt will fail in step (3) of the algorithm because *John* is not of the category Det. Backtracking requires choice points to be marked (e.g. stored in a data structure), to be able to cancel everything up to the last choice point in the event of a failure. This is not necessary in the case of programming with Prolog language, where the backtracking mechanism is natively implemented.

As backtracking has a non-negligible calculation cost, some researchers have explored the possibility of a deterministic strategy. In fact, the intuitive observation of the human parsing process shows us that the resolution of ambiguity is not performed by a research of the set of rules applicable to the sentence in the course of processing. To do this, humans seem to use all the knowledge they have to choose the solution that seems most plausible. The arguments in favor of this type of strategies are the extreme speed of the human parser and the non-conscientization of different possibilities as humans can interpret sentences without realizing the existence of several possibilities. This intuitive observation has given birth to several algorithms including those by [MAR 80] and [SAB 83].

Another control strategy, which is called breadth-first search, is to explore all possible rules at all points in the parsing. It is a horizontal exploration of the search space that allows us to identify all possible parses of a sentence in the case of a syntactic ambiguity. Such a strategy allows the *à posteriori* selection of the best parse according to syntactic, semantic, discursive, etc. criteria. From a cognitive point of view, such a strategy is not relevant, because humans do not seem to consider all possible analyses of a sentence at the same time. With respect to the computational point of view, this strategy is very costly in terms of space (memory) and parsing time. This is particularly true with actual applications where the grammar size is quite large, and sometimes it makes possible an infinite number of parses for a given sentence.

The problem of the left recursion is very often cited in the literature as the main limit of top-down algorithms, because of the rules of the form: A → A α will cause this algorithm to loop to infinity. To illustrate this problem, let us take the micro-grammar of the Figure 4.73.

S → NP VP

NP → NP PP

...

Figure 4.73. *Micro-grammar with a left recursion*

To analyze any sentence with the grammar of the Figure 4.73, the algorithm will first try the first rule S → NP VP and then it will try to develop the noun phrase with the rule NP → NP PP and it, therefore, falls in an infinite loop, as illustrated in Figure 4.74.

To resolve this problem, we could perform transformations on the form of the rules, but this will not provide linguistically relevant parse trees (see [GRU 95] for more details on this issue).

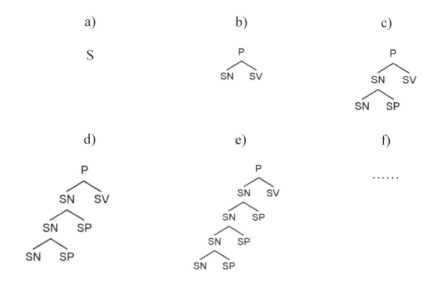

Figure 4.74. *Left recursion with a top-down algorithm*

4.4.4. *Bottom-up approach*

Unlike top-down algorithms, bottom-up algorithms begin with the words and then go up gradually to achieve a parse of the entire sentence, if it is possible. It consists of searching for the words of the sentence and then combining them together in higher order structures by using the grammar rules until the arrival at the axiom. The stack in the bottom-up algorithms contains the symbols that have already been parsed.

The simplest bottom-up algorithm is called shift-reduce. As indicated by its name, this algorithm is based on two operations:

– *shift*: allow the algorithm to move to the next word;

– *reduction*: try to combine the constituents already found in higher constituents. This operation is possible thanks to a stack which retains the already found constituents.

Continuing with our sentence: *Le livre est intéressant* (The book is interesting) and the grammar of the Figure 4.70 we get the parsing steps presented in Figure 4.75.

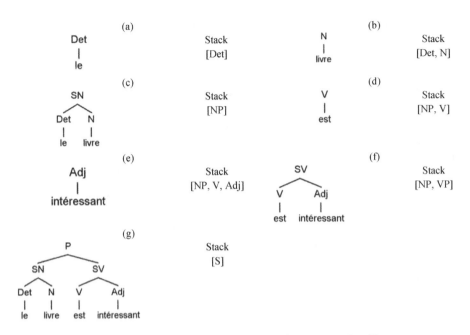

Figure 4.75. *Example of parsing with a bottom-up algorithm*

The steps of the shift-reduce algorithm are provided in Figure 4.76.

0.	Initialization: stack = []
1.	Either: shift
	• Select the next word of the sentence or the first word at the beginning.
	• Find the category of this word.
	• Place the category of this word at the top of the stack.
2.	Or: reduction
	• If the categories in the stack correspond to the right-hand side of one of the grammar rules, then
	– Remove the identified categories.
	– Add the symbol of the left-hand side of the rule in the stack.
3.	If there are no more words and if the stack = [S].
	• Then, success.
4.	Otherwise, go to step 2.

Figure 4.76. *Basic top-down algorithms*

The bottom-up algorithm provided in Figure 4.76 is based on the two operations that we have just presented. There are two alternatives of which only one is applied at a given point of the analysis.

If we go back to our example in Figure 4.75, the algorithm firstly begins the parsing with an empty stack. With the shift operation, it takes the first word of the sentence *the,* and it finds its grammatical category *det* that it puts in the stack (Step A). As there is no grammar rule in which the left-hand side is the non-terminal *det* (the unique symbol in the stack), the algorithm performs another shift: it finds the grammatical category *N* of the next word *book* and it adds it to the stack that now contains the two symbols *Det* and *N* (Step B). The algorithm finds a rule whose right-hand side is: *Det N.* It is the rule: NP → Det N. It performs a reduction which is to remove the two symbols that are located in the right-hand side of this rule of the stack and to replace them with the symbol which is located in its left-hand side: NP (Step C). Finally, in step (g), the two conditions of success are met: the algorithm has consumed all the input words and it has only the symbol *S* in the stack.

It is probably necessary to note finally that the top-down approach includes several commonly used parsing algorithms, such as CYK, EARLY [EAR 70] and GRL [*TOM* 86].

4.4.5. *Mixed approach: left-corner*

Historically attributed to [IRO 61, GRI 65] and [ROS 70], the left-corner approach is a hybrid approach that combines the methods of two approaches that we have just seen: the bottom-up approach and the top-down approach. This is performed in order to increase the efficiency of the parsing by reducing the search space.

The motivation behind the creation of this algorithm can be explained with the following example. Suppose that our algorithm has the fragment of grammar of the Figure 4.77 and that we have to parse the French interrogative sentence: *vient-il* (is he coming)?

```
S → NP VP
S → VP NP
NP → Det N
NP → NP
NP → PPS
VP → VI
.....
```

Figure 4.77. *CFG Grammar*

To analyze the input sentence, a top-down algorithm will perform the steps shown in Figure 4.78.

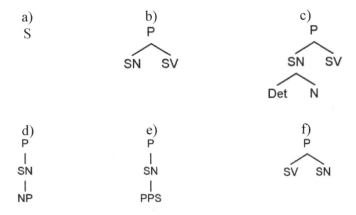

Figure 4.78. *Repeated backtracking with a top-down algorithm*

As we see in Figure 4.78, the algorithm starts to construct the tree from its root S. According to the grammar, a sentence can have two forms, thus the algorithm starts with the first (S → NP VP) (Step B). Then, on three occasions, it tries to find the noun phrase (Steps C, D and E) without success. In that case, due to the lack of other opportunities, it backtracks and tries the second form of the sentence (S → VP NP) which will eventually lead to the correct parse. In other words, to find the parse, the algorithm has performed four backtracks and this in a mini-grammar, then we can imagine the cumbersome nature of the research with a grammar of a large size.

Intuitively, we can say that the bottom-up approach is not cognitively relevant, because humans do not seem to examine all possibilities mechanically. In fact, a normally constituted human would move immediately to the second form of the sentence (S → VP NP), simply because the noun phrase cannot begin with a verb. This idea is the basis of the left-corner algorithm which uses precompiled tables of words or of grammatical categories that can appear at the beginning of a constituent. If we go back to our grammar of the Figure 4.70, a table of category and of lexicon can have the form provided in Table 4.10.

Category	Left-corner (categories)	Left-corner (lexicon)
S	Det, NP	the, my, John
NP	Det, NP	the, my, John
VP	V	is, loves, descends

Table 4.10. *Table of left-corners of the grammar of the Figure 4.70*

As we can see in Table 4.10, two types of information can be stored in the table: grammatical categories or lexical units. Depending on the size of the vocabulary, one of these choices is appropriate. In fact, in the case of a very large vocabulary as in the automatic translation applications of open texts, the compilation of a vocabulary with tens of thousands of words could cause memory problems for the system. In this case, we prefer the grammatical categories, in spite of the cost in terms of calculation.

Obviously, a rule which passes the lexical filter does not necessarily apply successfully, but this simply serves to reduce the search space and therefore increase the effectiveness of the algorithm.

A left-corner parser recognizes the left corner of a rule in a bottom-up fashion and predicts the rest of the symbols in the right-hand side of a rule in a top-down fashion. It includes three principal operations: shift, prediction and attachment.

Similar to the operation of the same name that we have just seen at the bottom-up algorithm, shift is to identify the category of the next word in the sentence (or the first word at the beginning) and to place in the stack the category which has been found. Prediction is to predict a constituent from its siblings. For example, suppose our stack has an NP at its top which corresponds to an already found noun phrase and that our grammar contains the rule (S $\rightarrow$ NP VP), then the NP will be replaced with S/VP. This means that the algorithm has predicted an S and that it must find the VP for the parsing of S to be successful. Attachment, meanwhile, is performed in two forms. The first is to consider a predicted constituent as already found when we find its constituents. For example, when we arrive at identifying an already predicted VP, the stack could have a form like [VP, S/VP]. Logically, the previous stack will be reduced to the form [S]. The second form is to combine the predictions that intersect. Thus, when we need to find an NP (complement) to complete a VP and if the latter is in turn necessary to complete a S, then we can deduce that we need an NP to complete S. The stack will be updated in the following manner: [VP/NP, S/VP] => [S/NP]. Note that a predicted element without having found any of its constituents will be noted as X/.

A possible form of the left-corner approach is provided in Figure 4.79. In fact, as the left-corner approach is a general strategy, it has been applied to various parsing algorithms such as GLC [NED 93], RTN [KUR 03] or tabular parsing [MOO 04].

Continuing with our example: *Le livre est intéressant* (The book is interesting) analyzed with the grammar 4.70 and the left-corner algorithm of the Figure 4.79, we get the steps presented in Figure 4.80. First of all, the algorithm predicts in a top-down approach that it is parsing a sentence and that is why it places the symbol S/ in the stack (Step A). In Step (b), it performs a shift by analyzing the left-corner of the sentence, the determiner The (*Le*), of which it finds the category (Det) that it places in the stack. With the rule (NP $\rightarrow$ Det N), it predicts that there must be a noun N to complete the already initiated NP and the stack is updated to reflect the new waiting expectation. Next, in step (c), after having found the N, it adds it to the stack. In step (d), it attaches the two constituents of the found NP. In step (e), it predicts a VP from the rule (S → NP VP) and the stack becomes [S/VP]. In other words, it announces that it has already found an NP to complete the predicted S and that it requires the VP. In step (h), it performs an attachment and then it declares the parsing successful; as the stack contains the symbol S and as all the words in the sentence have been consumed.

1.   Initialization: predict the initial stack symbol = [S/].
2.   Either: shift
    •   Select the next word of the sentence or the first word at the beginning.
    •   Find the grammatical category of this word.
    •   Place the category of this word at the top of the stack.
3.   Or: prediction
    •   Use the grammar rules to find the brother nodes of the found symbols and update the stack in the following way: if there is a rule in the form X → Y Z and if we have already found the symbol Y, we add X/Z at the top of the stack.
4.   Or: attachment
    •   If we find the non-terminal symbol(s) necessary to complete a predicted constituent, then we change the stack according to the diagram: [X/Y Y] => [X].
    •   If we predict a constituent and then we find another constituent which could be its child, then we can assume that we are on the right path to find the constituent. In practice, if we have the rules:
       X → Y Z and Z → A T and if the stack has the form: [X/Y, Y/T], it becomes [X/T].
5.   If there are no more words and if the stack = [S].
    •   Then, success.
6.   Otherwise, go to step 2.

Figure 4.79. *Left-corner algorithm*

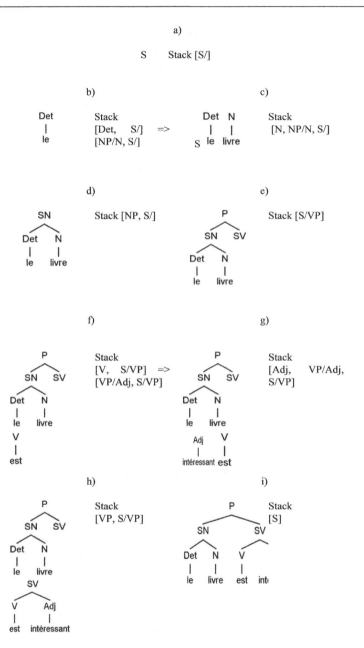

Figure 4.80. *Example of parsing with the left-corner algorithm*

In addition, several studies in computational psycholinguistics have shown that the top and top-down algorithms present difficulties as regards the processing of cases of left-branching and right-branching, respectively [JOH 83, ABN 91b, RES 92a]. This is due to the need for storage of intermediate information before being able to analyze the entire sentence. A summary of the spaces required by the three types of algorithms is provided in Table 4.11.

Strategy	Required space		
	Left	Auto	Right
Bottom	$O(n)$	$O(n)$	$O(1)$
Top	$O(1)$	$O(n)$	$O(n)$
Left-corner	$O(1)$	$O(n)$	$O(1)$

Table 4.11. *Summary of spaces required by the three parsing approaches [RES 92a]*

First of all, note that n in Table 4.11 represents the width of the trees in terms of nodes and that $O(1)$ represents a fixed processing time. Just like humans, the three strategies require more space to process the self-embedding structures regardless of their embedding degree. By contrast, we find an asymmetry in the processing of sentences with left-branching and right-branching for bottom and bottom-up algorithms. However, humans do not seem to process these two forms asymmetrically. This is considered as an argument in favor of the cognitive plausibility of the left-corner approach.

The left-corner approach cannot prevent the activation of irrelevant rules with respect to the context, particularly in the case of homography where two words with the same graphic form correspond to different grammatical categories such as the words *plant* and *use* which belong to both categories: verb and noun. In addition, the rules which do not have a left-corner, because their right-hand side is null (e.g. A $\rightarrow \phi$), require a particular processing. In contrast to the top-down approach, the left-corner can parse the left recursive rules to the left without looping to infinity, particularly thanks to lexical filtering.

4.4.6. *Tabular parsing (chart)*

Historically, the authorship of tabular parsing, which is commonly called *chart parsing*, is attributed to Martin Kay [KAY 67]. It is based on a dynamic programming approach, in which the main idea is the storage of the partial results found in the intermediate steps in order to reuse them in a subsequent step. The algorithm parses the same piece only once, which makes it more effective. Tabular parsing can be combined with the three parsing approaches that we have just seen, the top-down approach, the bottom-up approach and the corner-left approach. Similarly, it can use breadth-first or depth-first search strategies.

Tabular parsing is often compared to similar approaches, including the Earley algorithm and the Cocke–Younger–Kasami algorithm (CYK) [EAR 70, KAS 65, *YOU* 67]. Given the importance of this type of algorithm, several books have devoted large sections to it with detailed presentations including [VIN 83, GAZ 89, BLA 09, COV 94].

To store intermediate parses, it uses a data structure called *chart*. It is a set of nodes linked by edges. Each word of the input sentence is surrounded by two nodes that are marked by numbers. The edges at their turn mark the constituents. Let us take a look at the table of Figure 4.81 as an example.

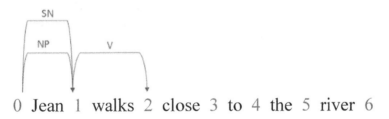

Figure 4.81. *Table of an incomplete parsing*

The table presented in Figure 4.81 contains numbers between 0 and 6 which surround the words of the sentence from the beginning to the end. The edges have the function to indicate the structures which have already been recognized by the parser. In our case, two edges mark a proper noun and an

NP between 0 and 1 and an edge marking a verb between 1 and 2. As for the progress of the parsing, the parser adds new edges to the table to account for identified constituents. Note that in the case of an ambiguity of morphological or syntactic category, the table contains two edges which connect the nodes which surround the ambiguous word with two different labels.

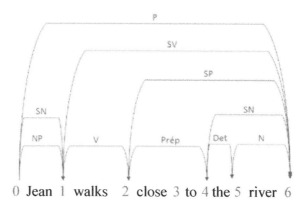

0 Jean 1 walks 2 close 3 to 4 the 5 river 6

Figure 4.82. *Table of a complete parsing of a sentence*

At the stage of parsing presented in Figure 4.82, the algorithm stops with success because, on the one hand, all possible edges between 0 and 6 are filled with an edge with the category *S* and, on the other hand, all the words in the input sentence had been analyzed. We also note that some elements are marked at several levels. For example, *John* is both a proper noun NP and a noun phrase NP and it is also a part of the sentence *S*.

The table in Figure 4.82 is a particular case that we commonly call "passive charts", because they do not realize the previously completed analyses that are partial or complete.

Active charts are the other type of tables which consists of realizing the state of parsing in a more comprehensive manner, i.e. what has been performed, as well as planned *projects* or steps. Concretely, an active chart contains active edges, i.e. edges whose requirements are only partially satisfied, as well as an agenda that defines the priority of task execution.

Active edges or dot rules are ordinary rewrite rules, but with an additional dot in their right-hand side that separates what has been analyzed from what has not been analyzed, or from what we call the rest. For example, the rule (S → NP • VP) indicates a situation where a noun phrase was found and where it requires a verb phrase to complete the parsing of the sentence.

If we go back to our example, *Le livre est intéressant* (The book is interesting) and the grammar of Figure 4.70, we get the active chart of the Figure 4.83.

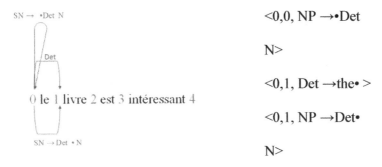

$<0,0, NP \rightarrow$•Det

N>

$<0,1, Det \rightarrow$the• >

$<0,1, NP \rightarrow$Det•

N>

Figure 4.83. *Partial active chart*

As we can see in Figure 4.83, we have two equivalent representations of the parse table with two active edges and a passive edge. The interpretation of this table is as follows. First of all, the parser predicts a constituent of NP type without recognizing any element (the dot is located at the right end of the right-hand side of the rule) and then after having identified a Det (marked by a passive edge), it moves the dot to the right of the identified constituent in the new initiated active edge. The question that now arises is how to combine passive edges with active edges to account for the progress of the parsing. To do this, tabular analyzers have a fundamental rule.

The intuition behind the fundamental rule is very simple. As an active edge expresses an expectation of specific constituents in a given context, the state of this expectation must be updated with the progress of the parsing and the discovery of one of these constituents that are expressed by the addition of a passive edge. This modification is performed simply by the advancement of the dot to the right of the symbol found in the rewrite rule. A more formal presentation of this rule is provided in Figure 4.84.

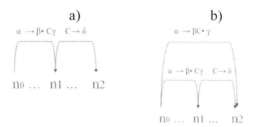

Figure 4.84. *Diagram of the first fundamental rule*

Let us take a concrete example: the sentence *Le livre est intéressant* (The book is interesting) analyzed with the grammar of the Figure 4.70 (Figure 4.85).

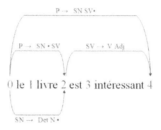

Figure 4.85. *Example of application of the fundamental rule*

The example presented in Figure 4.85 contains a labeled active edge (S →NP • VP). This means that the parser is expecting a VP. The following passive edge indicates that we have already found the VP initiated by the active edge. The fundamental rule allows us to combine the two edges and to advance the state of parsing considering that the sentence S has been fully parsed by generating an active edge labeled with (S →NP VP •).

To operate normally, a parser with an active chart must use an agenda. It is a data structure that allows us to plan the actions. In other words, the agenda of a tabular analyzer plays a similar role to the stacks that we have just seen: to memorize the actions and to establish priorities[9].

---

9 Another data structure, which is called queue, can also be used for the implementation of agendas. In this case, the insertion of new elements is performed at the end leading to a depth-first strategy (see [BLA 09] for more details).

In sum, tabular parsing, by storing partial results, avoids repeating the same parsing several times by performing backtrackings (as is the case with the simple algorithms that we have seen). Similarly, we have seen that this approach has an effective mechanism for processing ambiguities. It is probably necessary to mention that parsers of this type do not have a problem with the left recursion that is either with a top-down or a bottom-up strategy.

However, the addition of predictions to the parse table, the main advantage of tabular analyzers, has a few disadvantages. In fact, with either a top-down or a bottom-up strategy, this could mislead the parsing with incorrect predictions.

---

1. Initialize the table and the agenda

2. Repeat until the agenda is empty

    a. Take the first edge of the agenda

    b. Add the edge to the table

    c. Use the fundamental rule to combine this edge with the other edges of the table.

    d. The new edges of the previous step must be added to the agenda.

    e. Add new active edges (expectations) to the agenda based on existing edges and grammar rules.

3. If the table contains a labeled passive edge with S, then success, otherwise, failure.

Figure 4.86. *Tabular parsing algorithm with a bottom-up approach*

4.4.7. *Probabilistic parsing*

The idea behind statistical parsing is to combine the discrete symbolic information with the statistical information of a continuing nature. In practice, this is to enrich the existing linguistic formalisms with statistics obtained from corpora related to application fields [CHA 93a, BOD 95, ABN 96]. The way in which the Probabilistic CFG (PCFG) extends the CFG is often compared to the extension of regular grammars by the HMMs that we have seen in Chapter 2. The objectives of these different approaches can be summarized by ambiguity resolution, as well as the limitation of the search space of classical parsing algorithms. Studies in this context have been performed in two complementary axes: the definition of statistical formalisms, as well as the proposition of algorithms.

The definition of probabilistic formalisms, in the general case, is to propose probabilistic versions of existing formalisms. We can mention in this framework the studies on probabilistic CFGs [JEL 92], stochastic LTAG [SCH 92, RES 92b], stochastic LFG [BOD 00, BOD 98], etc. Sometimes, probabilistic analysis is proposed for the resolution of specific problems, such as processing of syntactic discontinuities [LEV 05] or ambiguity resolution [TOU 05]. The main problem addressed by these studies is the hierarchization of probabilities according to linguistic units. Thus, we distinguish between internal probabilities, within the basic linguistic unit such as the phrase and the chunk and inter-unit external probabilities.

The other axis consists of the proposition of algorithms for probabilistic analysis. In this context, we can mention the studies by [MAG 94, BOD 95, GOO 98], etc. The main issues targeted on this axis are the obtaining of the most probable parse tree (ambiguity resolution) effectively, the learning automation of PCFG, as well as the reduction of the size of the learning corpora.

The probabilistic analysis is a method which has several benefits for the processing procedure, but its practical realization is generally difficult, particularly in the context of deep parsing systems, especially because of the need for a large quantity of correctly pre-parsed data.

From a formal point of view, a PCFG grammar can be defined as $G = (V_N, V_T, S, R, \Pi)$. The only difference compared to the definition of the CFG that we have seen is the element Π which is the set of probabilities associated with production rules with: $\pi(\alpha \rightarrow \beta) \geq 0$ where $\alpha \rightarrow \beta \in R$.

As we note in Figure 4.87, the sum of probabilities of the rules which share the same non-terminal symbol in their left-hand side is equal to one. More formally, we can say that the PCFG obey the property formulated by the following equation: $\forall \ \alpha \in VN, \sum_i \pi(\alpha \rightarrow \beta i) = 1$. For example, if we take the case of rules of NP: $\pi(NP \rightarrow Det\ N) + \pi(NP \rightarrow PN) + \pi(NP \rightarrow Pron) = 0.27 + 0.63 + 0.1 = 1$.

S → NP VP	0.7
S → VP	0.3
NP → Det N	0.27
NP → NP	0.63
NP → Pron	0.1
VP → V Adj	0.5
VP → V	0.3
VP → NP V	0.2
Det → my	0.45
Det → the	0.55
NP → John	1.0
V → book	0.1
V → loves	0.3
V → is	0.5
V → descends	0.1
N → book	0.35
N → notebook	0.40
N → program	0.25
Adj → interesting	0.6
Adj → expensive	0.4
Pron → the	0.3
Pron → the	0.3
Pron → the	0.4

Figure 4.87. *Example of a probabilistic context-free grammar for a fragment of French*

To calculate the probability of an parsing tree, simply multiply the probabilities of the rules used in its derivation, as we assume that the probabilities of the rules are independent. Thus, the probabilities of trees a_1 and a_2 in Figure 4.88 are calculated in the following manner from the grammar of the Figure 4.87:

$p(a_1) = \pi(p \rightarrow NP\ VP)\ \pi(NP \rightarrow NP)\ \pi(NP \rightarrow John)\ \pi(VP \rightarrow V)\ \pi(V \rightarrow descends) = 0.7 * 0.63 * 1 * 0.3 * 0.1 = 0.0132$.

$p(a_2) = \pi(p \rightarrow NP\ VP)\ \pi(NP \rightarrow Det\ N)\ \pi(Det \rightarrow the)\ \pi(N \rightarrow notebook)\ \pi(VP \rightarrow V\ Adj)\ \pi(V \rightarrow is)\ \pi(Adj \rightarrow expensive) = 0.7 * 0.27 * 0.55 * 0.4 * 0.5 * 0.5 * 0.4 = 0.004158$.

As we note, logically, the larger the parsing tree is, the smaller its probability is.

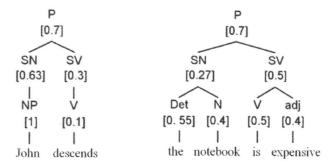

Figure 4.88. *Parsing tree for a sentence from the PCFG of the Figure 4.87*

In the case of the two sentences of Figure 4.88, each sentence is associated with a single parse, as there is no syntactic ambiguity. What is interesting about a probabilistic grammar is that it assigns a probability to all analyses or possible derivations of a syntactically ambiguous sentence, allowing us to resolve the ambiguity. More formally, suppose we have a sentence S which is likely to receive n different parses, the probability of this sentence can be calculated according to the equation [4.33]. Thus, for each derivation d member of the set of derivations of S, we have a probability associated with this derivation $\tau(S)$. The probability of a sentence is equal to the sum of the probabilities of its derivations.

$$p(S) = \sum_{\tau \in \tau(S)} P(A,S) \qquad [4.33]$$

Despite its benefits, PCFG suffers from several well-known limitations in the literature. In fact, according to the independence hypothesis, each rule is independent of other rules which constitute the parsing tree. Thus, the rule $S \rightarrow NP\ VP$ does not imply any limitation with respect to the form of the subject noun phrase. However, studies such as [FRA 99] have shown that in English in 91% of the cases, the subject NP is a pronoun. In addition, conjunction ambiguities as in [4.16a] are impossible to solve with a PCFG, since the two possible parses have the same ambiguity. To resolve these problems, several strategies have been proposed in the literature. The simplest strategy is to differentiate labels. For example, rather than having the category NP, we would have NP_Subject and NP_Object. A more linguistically motivated solution is to lexicalize the grammar and to associate probabilities to lexically anchored trees. This allows us to take into consideration several syntactic parsing levels.

In a practical manner, PCFGs are inferred generally from a syntactically annotated corpus, which is often called *TreeBank* (see Figure 4.89 for a describing this process). The inference algorithm explores the trees in the corpus and calculates the probabilities of each of the rules observed by using the [4.34].

$$P(\alpha \to \beta \,|\,\alpha) = \frac{c(\alpha \to \beta)}{c(\alpha)} \qquad\qquad [4.34]$$

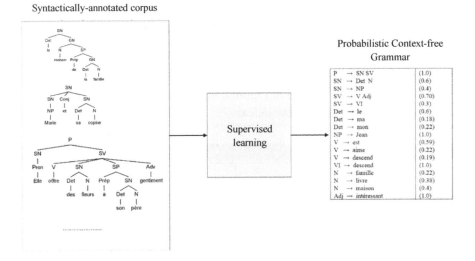

Figure 4.89. *Supervised learning of a PCFG*

Another approach is to infer the PCFG from a non-annotated corpus. Typically, we use the inside-outside algorithm whose principle is similar to the Baum-Welch algorithm for the estimation of the parameters of an HMM. As we might think, these studies are still at the research stage and their performance is still inferior to that of the supervised approach.

Several probabilistic parsing algorithms have been proposed, many of which are inspired by the tabular parsing approach described in section 4.4.6.

To find the best parse among the possible parses, algorithm such as A* or Viterbi, that we have presented in Chapter 2, are used [COL 99, KLE 03].

To give a concrete idea of the probabilistic syntactic parsing, let us take the CYK algorithm as an example. Named after its historical inventors, Coke, Yonger and Kasami, this algorithm is a tabular algorithm which follows a bottom-up approach. It requires grammars in Chomsky normal form, in which all rules above the preterminal level are binary. Effective, this algorithm has a processing time complexity which is equal to $O(n^3)$ and requires a space of $O(n^2)$. As it follows a dynamic programming approach, this algorithm searches for all partial parses for the sequences of equal length up to n words where n is the number of words of the input sentence. The probabilistic version of this algorithm uses the same approach, but it is distinguished by the association of a probability to each sequence produced by a rule (see [CHA 98] for an example). If we take as an example the French sentence *Le livre est intéressant* (The book is interesting) with the grammar 4.87, we get the following parse steps. The first step is to create a data structure which is called table or parsing triangle, as it has the shape of an inverted right triangle whose two catheti are formed by n cells where n is equal to the number of words of the input string. As we can see in Figure 4.90, each cell covers a substring with a beginning index and an ending index.

Le	livre	est	intéressant
|0,1|	|0,2|	|0,3|	|0,4|
	|1,2|	|1,3|	|1,4|
		|2,3|	|2,4|
			|3,4|

Figure 4.90. *General structure of the parse table of the CYK algorithm*

The parsing algorithm proceeds by filling the hypotenuse of the rectangle with the information on the morphological categories of input words introducing the rules of the form: preterminal → word (see Figure 4.91). Lexical ambiguities are retained to the extent that we add the different parsing possibilities of a word with their probabilities. For example, for the word *le* (*the*), we add the two possible rules in the grammar: Det → the 0.55 and Pron → the 0.3. Next, the algorithm searches if there are higher level structures which have the preterminal used to the right-hand side. For example, we find that it is possible to construct a noun phrase with the following rule: NP → Pron. The probability associated with this rule in the parse table is calculated as follows: π(NP → Pron) * π(Pron → the) = 0.1*0.3=0.03.

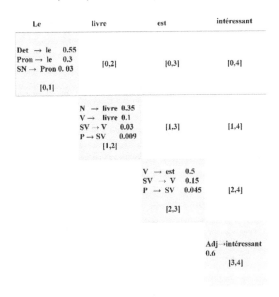

Figure 4.91. *The first step in the execution of the CYK algorithm*

In the second diagonal line, the algorithm searches for strings of two words by combining the representations which have already been constructed at the previous level, that of words (see Figure 4.92). For example, it finds that it is possible to construct a sentence with the rule S → NP VP [0, 2] whose probability is calculated in the following manner: π(S → NP VP) π(NP → Pron) π(VP → V) = 0.7*0.0006*0.051 = 0.000021. In a similar manner, the algorithm discovers the possibility of obtaining a sentence in the cell [2, 4]. The cell [1,3] remains empty because the string *livre est (book is)* does not correspond to any constituent.

In the third step, the algorithm finds the possibility to construct a sentence with the rule S → NP VP in the cell [0.3] (Figure 4.93).

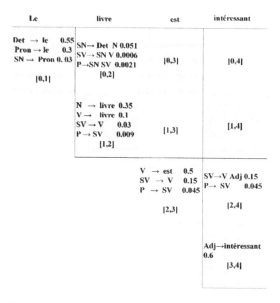

Figure 4.92. *The second step in the execution of the CYK algorithm*

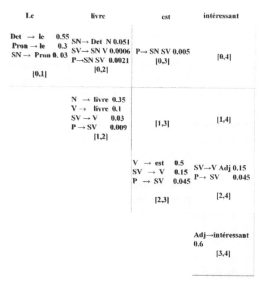

Figure 4.93. *The third step in the execution of the CYK algorithm*

Finally, the algorithm finds a rule whose left-hand side is *p* and which covers the entire sentence in the cell [0, 4] (Figure 4.94). This allows us to conclude that the string *Le livre est intéressant* (The book is interesting) is a legitimate sentence according to our grammar. In the case of ambiguity, i.e. two different rules with the same non-terminal symbol in their left-hand side and which cover exactly the same string, we keep only the most probable rule. This is the mechanism that allows the CYK algorithm to resolve the ambiguity.

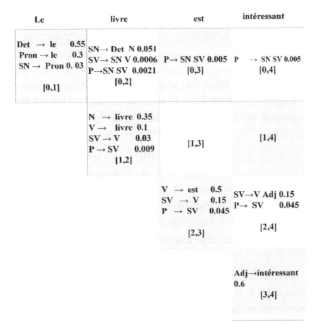

Figure 4.94. *The fourth step in the execution of the CYK algorithm*

To find the parse tree, we must enrich the rules with indexes to indicate the extent of each constituent. For example, the rule S → NP VP in the cell [0,4] will be marked in the following manner: S → NP [0,2] VP [2,4].

4.4.8. *Neural networks*

As we have seen in the sphere of speech, the first investigations on neural networks were presented in the 1940s with the studies by Walter Pitts and Warren McCulloch. After a series of partial successes and failures, neural networks have begun to occupy an important place in the landscape of artificial intelligence in the general sense of the term from the 1980s, particularly following the publication of the book *parallel distributed processing* by David Rumellheart and his collaborator [RUM 86] (see [JUT 94] and [JOD 94] for an introduction).

Today neural networks bring together an important family of statistical learning algorithms. Some of them, such as perceptrons, follow the supervised learning mode where the network learns by comparing the given input with the correct classification. Other networks adopt a non-supervised learning mode such as Kohonen maps or *self-organizing maps* (SOM) [KOH 82].

To be concrete, let us take the example of multilayer perceptrons which are a popular form of supervised neural networks. The architecture of this system is composed of three parts: the input layer, one or several hidden layers and the output layer. The input layer has the function of introducing the coded data to the network. For example, in the task of handwritten digit recognition, this layer can consist of n neurons where n is equal to the number of pixels of the images of each digit. Thus, with images of 28 × 28 pixels, we must have 784 neurons in the input layer. The hidden layer, in turn, constructs an internal representation within the network. Finally, the output layer corresponds to the classes of our system which, in the case of digit recognition, are ten (see network architecture in Figure 4.95). Each neuron in the network is connected to a subset of neurons in the network. Each connection has a weight whose value can change during the learning process. Neurons also have an activation function f which controls their outputs given their inputs. The input of a neuron is the weighted sum of neuron outputs of which it receives the output. The complexity of this function varies depending on the networks. The simplest function is called identity function, where the output of a network is equal to its net input: $f(x) = x$. In this case, we refer to a linear network. More complex activation functions are also used depending on applications such as the *step function* with $f(x) = 0$ if $x \leq 0$ and $f(x) = 1$ if $x > 0$ or the sigmoid function:

$$f(x) = \frac{1}{1 + e^{(-ax)}} .$$

There are two different but complementary forms of learning in a neural. The first one is to store diverse information in a compact form, such as in a Hopfield network [HOP 82]. The second method is to deduct forms of implicit rules that generalize the cases observed in the concrete examples provided in learning.

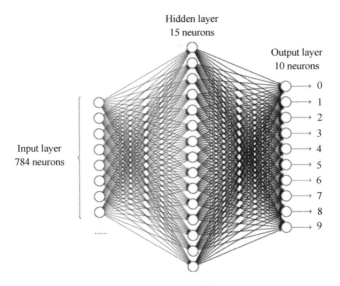

Figure 4.95. *Architecture of a neural network for handwritten digit recognition [NIE 14]*

The main limitation to applying networks such as the one in Figure 4.91 is the lack of consideration of context, since digits are recognized only based on their content. To take into account contextual information and temporal sequences, Elman has proposed a particular architecture of neural networks called recurring networks [ELM 90]. Unlike traditional networks, where the flow of information always goes forward: from the input to the output, *feedforward*, the recurrent networks allow for a return of information rearward. Inspired by the biological neural networks, this bidirectionality allows us to account for the context by memorizing implicitly one or several history stages (see Figure 4.96).

This type of network has been the subject of several studies for the construction of artificial neural networks which are capable of performing syntactic parsing effectively [JOD 93, HEN 94, GER 01] (see [HEN 10] for a review of these studies).

Input layer   Hidden layer   Output layer

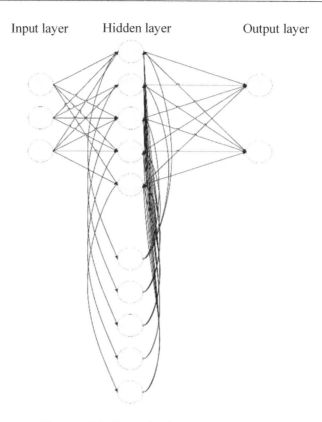

Figure 4.96. *Example of a recurring network*

Despite their interest, the results with neural networks were often inferior to those of statistical approaches. Even in the case of comparable performance, as for speech recognition, neural networks have been disadvantaged because of the high cost of learning. In fact, for actual applications learning required sometimes weeks, which means that researchers had a very low margin of error. A consequence of this cumbersome nature of learning was the inability to equip networks with the appropriate number of hidden layers to construct rich enough internal representations, because the addition of new internal layers dramatically increases learning time. To resolve this problem, a group of Canadian researchers including Geoffrey Hinton have proposed a new paradigm named deep learning [HIN 06, HIN 12] (see [BEN 09] for an introduction). This has opened the way for the exploration of new architectures of neural networks to solve a fairly large number of problems including the tricky problem of parsing.

Thus, several studies have focused on building syntactic parsers with this new type of neural networks. Typically, these studies focus on dependency formalisms [COL 11, ETS 13, CHE 14].

4.4.9. *Parsing algorithms for unification-based grammars*

Unification grammars, as we have seen, contain complex feature structures. We have also seen a simple way to process the additional information encoded by the features in the framework of DCG.

Let us go back to the tabular parsing algorithm to examine how it can be adapted to parse unification-based grammars. The principle is simple: in addition to the constraints on the grammatical categories imposed by grammar, the parser requires the unification of features associated with each example. It is therefore an addition of new constraints.

If we assume that the algorithm adopts a top-down approach, then it will first predict an active edge of the form:

$$\left\langle 0,0[\text{Cat} \quad \text{P}] \to \bullet \begin{bmatrix} \text{CAT} & \text{SN} \\ \text{Nb.} & \boxed{1} \end{bmatrix} \begin{bmatrix} \text{CAT} & \text{SV} \\ \text{Nb.} & \boxed{1} \end{bmatrix} \right\rangle$$

In the previous rule, the noun phrase and the verb phrase have a reentrant feature to ensure the agreement of the verb with its subject. Next, by processing the word *John*, it will add the following passive edge where we have voluntarily simplified it by considering that a proper noun is directly sufficient for a noun phrase:

$$\left\langle 0,1 \begin{bmatrix} \text{CAT} & \text{SN} \\ \text{Nb.} & \text{Sing} \end{bmatrix} \right\rangle \to \text{John} \bullet$$

With the fundamental rule, we advance the point to the right of the symbol.

$$\left\langle 0,1[\text{Cat} \quad \text{P}] \to \begin{bmatrix} \text{CAT} & \text{SN} \\ \text{Nb.} & \boxed{1} \end{bmatrix} \bullet \begin{bmatrix} \text{CAT} & \text{SV} \\ \text{Nb.} & \boxed{1} \end{bmatrix} \right\rangle$$

The rest of the sentence is processed in the same way until obtaining a complete parse: all words are covered under the category *S*. Note that, in addition, the algorithm must verify the unification of the reentrant features to ensure agreement.

4.4.10. *Robust parsing approaches*

A robust parser is a system that is able to provide a correct parsing tree even in the case of an incomplete, distorted or unexpected input. Several robust parsing techniques have been developed in the framework of studies on the spoken, as well as on the written language. In both cases, the objective is the use of parsing algorithms in actual conditions: recognition errors and speech extragrammaticalities for spoken language and typing errors and grammatical errors in written texts. The main techniques used in robust parsing consist of extensions of conventional parsing algorithms, in order to energize them and make them more appropriate to unexpected actual applications. In addition, some approaches are based on studies in fields such as information retrieval and document classification.

4.4.10.1. *Chunk parsing*

Inspired by the studies of [GEE 83] in psycholinguistics, [ABN 90, ABN 95] proposes a partial parsing approach based on the segment (chunk parsing).

In a chunk parser, parsing is divided into two completely distinct parts (unlike the traditional approaches in which the two steps are merged):

– segmentation: the conversion of the flow of words in a flow of segments;

– attachment: the attachment of the segments obtained in the previous phase within a global structure, which is the parse tree of the utterance. In regard to the previous part, this step is not mandatory or at least not systematic. Thus, a chunker can provide complete parsing trees and partial segments or only partial segments.

Cascaded parsing of syntactic structure (CASS) is a robust parser system based on chunks. This system has been developed by Steven Abney at the University of Tübingen in Germany. CASS uses a set of simple parsers that apply in cascade to construct a global syntactic representation of the utterance.

The input of CASS is the output of Church's tagger [CHU 8] which provides POS tags to the words, as well as the simple (non-recursive) noun phrases. Note that the processing rate of noun phrases is lower than that of POS tags. The processing of the pre-processed input in CASS is performed according to three steps: the chunk filter, the clause filter and the parse filter.

In turn, the chunk filter is based on two subfilters: the noun phrases filter and the chunk filter. The noun phrases filter is a module which uses regular expressions to assemble noun phrases based on the superficial analysis provided by Church's noun phrase recognizer. Similarly in this module, we correct the processing errors of noun phrases by the Church's module, such as those resulting from prenominal adjectives. In addition, the chunk filter also uses regular expressions to recognize the rest of the segments. Here is an example of the output of this module with the utterance: *In south Australia beds of Boulders were deposited.*

CS

[pp in [Np south Australia beds]]

[pp of[Np boulders]]

[Vp were deposited]

CS

As we can see, the system has committed an error of parsing (because of the tagger) of the first noun phrase *south Australia beds*.

The clause filter also consists of two subfilters: the raw filter and the corrected clause filter. The raw filter tries to recognize the boundaries of simple clauses, as well as marking the subject and the predicate of the clause. If it is unable to identify a single subject or predicate, this module identifies the type of error encountered, such as the existence of several verbal phrases or the absence of a subject (because of an ellipsis, for example), etc. In addition, the corrected clause filter is a module that tries to correct the errors identified by the previous module by applying specific patterns in each case. For example, the following is the pattern used for the correction of non-analyzed complementers: [pp X_p-time NP] ... VP $\rightarrow$ [$_{clause}$ X_c NP ... VP]. In cases where none of the patterns are applicable to the input, the system uses general heuristics that allow it to improve the analysis based on partial information (such as the existence of a noun phrase next to a verb phrase, only a verb phrase, etc.). Thus, after this step, the parse obtained for the utterance becomes as follows:

[$_{pp}$ in south Australia]

[$_{Subj}$ [Np beds]]

[$_{pp}$ of boulders]

[$_{Pred}$ [Vp were deposited]]

As we can see in the previous parse, the system has succeeded in correcting the error of parsing in the first phrase.

Unlike the previous modules, the parsing filter is based on recursive rules (not regular expressions). The main function of this module is to assemble recursive structures by attaching the nodes to each other according to the nature of the heads of these structures and the grammatical constraints on their assembly. For example, a segment Y can be attached to a segment X only if the head of X can have Y as an argument or modifier.

The results of CASS have shown that it is both robust enough and very fast for the processing of written corpora. These results are mainly due to the architecture of the system which is to apply different parsing levels in cascade with rules and patterns that allow us to correct the errors made in the previous steps.

Different approaches similar to Abney's approach were proposed including that by [AIT 97, GRE 99] which is based on techniques of FSAs and that of supertagging proposed by [CHE 99, SRI 99], in the framework of LTAG formalism.

4.4.10.2. *Selective approaches*

Selective approaches, which are sometimes called *island-driven parsing*, consist of analyzing only the parties considered relevant or non-noisy by the received utterance. These approaches are supported by simple observations on human language and speech processing which is characterized by the variation of the degree of attention. From a computational point of view, it often refers to equipping the parsing algorithm with a filter which allows us, according to a number of constraints, to ignore one or more words or the non-parsable segments. In fact, when analyzing the oral language, the input of the parser is the output of the module for speech recognition which contains a variable error rate depending on the circumstances (the amount of noise, the speaker's clarity of pronunciation, the richness of vocabulary, etc.).

Different degrees of selectivity have been used in the literature. This varies between fairly similar approaches to the parsing based on keywords such as [LUZ 87, ROU 00] to more reasonable coverage approaches such as the algorithm GLR* by [LAV 97] or the different implementations of semantic grammars at Carnegie Mellon University, as well as in other universities, [MAY 95, MIN 96, GAV 00, BOU 02].

Contrary to what some researchers in the field believe, selective approaches are not necessarily synonymous with information loss or shallow parsing. In fact, a well-designed selective strategy can be added to any parsing system without affecting its depth of analysis (see [KUR 03] for a more thorough discussion). The only drawback of these approaches is that they increase the computational complexity of algorithms to which they are added. For example, Wang [WAN 01] describes a chart parsing algorithm which is equipped with a selective strategy (for the parsing with a semantic grammar equivalent to CFG) whose complexity is $O(b^4)$ instead of $O(b^3)$, as is the case of several conventional algorithms for the CFG[10].

4.4.10.3. *Parsing of extragrammaticalities of the oral language*

The presence of extragrammaticalities in the oral language makes the use of conventional algorithms such as those we have seen impossible. For example, during our spoken language study done on the Trains Corpus, a corpus of spontaneous oral conversations in American English, we have found that filled pauses or hesitations constitute approximately 7% of the words of this corpus and approximately 66,26% of the utterances contain hesitations, in addition to a significant number of more complex phenomena,

---

10 This information is indirectly mentioned in Wang's article, but it has been explicitly provided during the oral presentation of this article at the Eurospeech conference 2001 in Aalborg, Denmark.

such as repetitions, self-corrections and false starts. To overcome the difficulties of parsing related to extragrammaticalities of the oral language, several sources of knowledge have been used in the literature for the processing of extragrammaticalities, including structural information, morphosyntactic information, prosodic information.

Structural information focuses on the identity of each word and those of the words which follow them[11]. The advantage of this information is its reliability and simplicity of use, but its use is generally limited to the detection of repetitions with the help of patterns [SHR 94, HEE 97].

Morphosyntactic information essentially concerns the morphosyntactic categories of words or chunks and their possible successions. For example, the succession of two determiners is deemed extragrammatical and therefore the case is processed as an self-correction. Some systems have used more complex rules to model cases involving phrases. In these kinds of cases, conventional parsers have been developed to perform this task. These rules have generally been implemented as syntactic metarules in a post-processing module [COR 97, MCK 98, COR 99, KUR 02].

Prosodic information relies on a set of sources of various kinds, such as unfilled pauses and the melodic contour, which have been used to segment the input in syntactic constituents and therefore locate the center of the extragrammaticality in the utterance [NAK 94, LIC 94].

4.4.11. *Generation algorithms*

As we have seen with the parsing algorithms, the input is a string of words (typically a sentence) and the expected output of the algorithm is a description of syntactic properties of this input string as tree or a dependency graph. Then comes the role of semantic knowledge to provide an interpretation of the sentence. In the case of generation, it is the reverse process, because we must first consider what we would like to express in terms of semantic content and then we produce the syntactic structure from which we find the surface string. We can also consider a generation at the scale of an entire text where consideration of discursive knowledge is indispensable, particularly when taking into account the connections

---

11 This means that the system verifies if two words are identical or not, regardless of their respective morphological categories.

between the sentences. We refer to [DAN 00, REI 10] for a general introduction to the issues of syntactic and semantic generation. With regard to the syntactic aspects of generation, we often use top-down algorithms sometimes with a tabular algorithm. Unification is also used to control the forms of the sentences generated in many approaches including [GER 90, EST 90, REI 04].

Bibliography

[ABE 93] ABEILLÉ A., *Les nouvelles syntaxes: grammaires d'unification et analyse du français*, Armand Colin, Paris, 1993.

[ABE 00] ABEILLÉ A., CANDITO M.-H., "FTAG: a lexicalized tree adjoining grammar for French", in ABEILLÉ A., RAMBOW O. (eds), *Tree Adjoining Grammars*, CSLI, Stanford, CA, 2000.

[ABE 03] ABEILLÉ A., CLÉMENT L., TOUSSENEL F., "Building a treebank for French", in ABEILLÉ A. (ed.), *Treebanks*, Kluwer, Dordrecht, 2003.

[ABN 87] ABNEY S., The English noun phrase in its sentential aspect, PhD Dissertation, MIT, 1987.

[ABN 90] ABNEY S., "Rapid incremental parsing with repair", *6th New OED Conference: Electronic Text Research*, University of Waterloo, Waterloo, Ontario, pp. 1–9, October 1990.

[ABN 91a] ABNEY S., "Parsing by chunks", in BERWICK R., ABNEY S., TENNY C. (eds), *Principle Based Parsing*, Kluwer Academic Publishers, Dordrecht, 1991.

[ABN 91b] ABNEY S., JOHNSON M., "Memory requirement and local ambiguities of parsing strategies", *Journal of Psycholinguistic Research*, vol. 20, no. 3, pp. 233–250, 1991.

[ABN 95] ABNEY S., "Chunks and dependencies: Bringing processing evidence to bear on syntax", in *Computational Linguistics and the Foundations of Linguistic Theory*, CSLI, available at: http://www.vinartus.net/spa/91i.pdf, 1995.

[ABN 96] ABNEY S., "Statistical methods and linguistics", in KLAVANS J., RESNIK P. (eds), *The Balancing Act*, The MIT Press, Cambridge, MA, 1996.

[ABR 84] ABRAMSON H., "Definite clause translation grammars", *The International Symposium on Logic Programming*, Atlantic City, New Jersey, pp. 233–240, 6–9 February 1984.

[AHO 88] AHO A.V., SETHI R., ULLMAN J., *Compilers: Principles, Techniques and Tools*, Addison-Wesley, Reading, MA, 1988.

[AIT 84] AIT-KACI H., A lattice theoretic approach to computation based on a calculus of partially ordered type structures (property inheritance, semantic nets, graph unification), PhD Thesis, University of Pennsylvania, 1984.

[AIT 97] AIT-MOKHTAR S., CHANOD J.-P., "Incremental finite-state parsing", *Proceedings of the 5th International Conference on Applied Natural Language Processing (ANLP)*, Washington, DC, pp. 72–79, 1997.

[ALD 07] ALDERETE J.D., FRISCH S.A., "Dissimilation in grammar and the lexicon", in DE LACY P. (ed.), *Cambridge Handbook of Phonology*, Cambridge University Press, 2007.

[ALF 89] ALFOZAN A., Assimilation in classical Arabic: a phonological study, PhD Dissertation, University of Glasgow, 1989.

[ANT 91] ANTWORTH E.L., "Introduction to two-level phonology", *Electronic Document, Notes on Linguistics No. 53*, pp. 4–18, available at: http://www-01.sil.org/pckimmo/two-level_phon.html, 1991.

[ANT 94] ANTOINE J.-Y., Coopération-syntaxe sémantique pour la compréhension de la parole spontanée, Thesis, Institut National Polytechnique de Grenoble, 1994.

[ANT 01] ANTOINE J.-Y., GOULIAN J., "Word order variations and spoken man-machine dialogue in French: a corpus analysis on the ATIS domain", *Corpus Linguistics'2001*, Lancaster, UK, pp. 22–29, 2001.

[ANT 02] ANTOINE J.-Y., LETELLIER-ZARSHENAS S., NICOLAS P. *et al.*, "Corpus OTG et ECOLE_MASSY: vers la constitution d'une collection de corpus francophones de dialogue oral diffusés librement", *Actes TALN'2002*, Nancy, France, pp. 319–324, June 2002.

[ARA 02] ARAUJO L., "Part-of-speech tagging with evolutionary algorithms", in GELBUKH A. (ed.), *Computational Linguistics and Intelligent Text Processing*, Springer, Heidelberg, 2002.

[ARC 84] ARCHANGELLI D., On the Nature of Phonological Representations: Underspecification Theory, Thesis, Brandeis University, January 1984.

[ARC 88] ARCHANGELLI D., *Underspecification in Yawelmani Phonology and Morphology (Outstanding Dissertations in Linguistics)*, Garland Publishing, New York, 1988.

[ASM 14] ASMUSSEN J., "Survey of POS taggers", Technical Report, Society for Danish Language and Littérature, DSL, available at: korpus.dsl.dk/clarin/corpus-doc/pos-survey.pdf, 5 March 2014.

[ATW 08] ATWELL E., Corpus linguistics and language learning: bootstrapping linguistic knowledge and resources from text, PhD Dissertation, University of Leeds, 2008.

[BAC 59] BACKUS J.W., "The syntax and semantics of the proposed", *International Algebraic Language of the Zurich ACM-GAMM Conference, International Conference on Information Processing*, UNESCO, pp. 125–132, 1959.

[BAK 75a] BAKER J.K., "The Dragon system – An overview", *IEEE Trans, Acoust. Speech Signal Process*, pp. 24–29, 1975.

[BAK 75b] BAKER J.K., Stochastic modeling as a means of automatic speech recognition, PhD Thesis, Carnegie-Mellon University, 1975.

[BEE 03] BEESLEY K.R., *Finite State Morphology*, CSLI Publications, Stanford, 2003.

[BEN 02] BENDER E.M., FLICKINGER D., "Oepen Stephan, the grammar matrix: an open-source starter-kit for the rapid development of cross-linguistically consistent broad-coverage precision grammars", *Proceedings of the Workshop on Grammar Engineering and Evaluation at the 19th International Conference on Computational Linguistics*, Taipei, Taiwan, pp. 8–14, 2002.

[BEN 09] BENGIO Y., "Learning deep architectures for AI", *Foundations and Trends in Machine Learning*, vol. 2, no. 1, pp. 1–127, 2009.

[BEN 13] BEN ALI B., JARRAY F., "Genetic approach for Arabic part of speech tagging", *International Journal on Natural Language Computing*, vol. 2, no. 3, June pp. 1–13, 2013.

[BES 95] BESSAC M., CAELEN J., "Analyses pragmatiques, prosodiques et lexicales d'un corpus de dialogue oral homme-machine", *JADT'95*, Roma, Italy, pp. 363–370, 1995.

[BIR 03] BIRD S., "Phonology", in MITKOV R. (ed.), *Oxford Handbook of Computational Linguistics*, Oxford University Press, 2003.

[BIR 09] BIRD S., KLEIN E., LOPER E., *Natural Language Processing with Python*, O'Reilly, 2009.

[BLA 87] BLANCHE-BENVENISTE C., JEANJEAN C., *Le Français parlé: transcription et édition*, Didier Erudition, Paris, 1987.

[BLA 95] BLACHE P., Une introduction à HPSG, available at: http://aune.lpl.univ-aix.fr/~blache/publis.html, 1995.

[BLA 00] BLACK A.W., Speech Synthesis in Festival: a practical course on making computers talk, Language Technologies Institute, Carnegie Mellon University, available at: http://festvox.org/festtut/notes/festtut_toc.html, 2000.

[BLA 09] BLACKBURN P., STRIEGNITZ K., *Natural Language Processing Techniques in Prolog*, O'Reilly Media, 2009.

[BLO 48] BLOCH B., "A set of postulates for phonemic analysis", *Language*, vol. 24, pp. 3–46, 1948.

[BLU 03] BLUTNER R., ZEEVAT H. (eds), *Optimality Theory and Pragmatics*, Palgrave Macmillan, London, 2003.

[BOD 95] BOD R., Enriching linguistics with statistics: performance models of natural language, PhD Dissertation, University of Amsterdam, 1995.

[BOD 98] BOD R., KAPLAN R., "A probabilistic corpus-driven model for lexical functional analysis", *Proceedings COLING-ACL '98*, Montreal, Canada, 1998.

[BOD 00] BOD R., "An empirical evaluation of LFG-DOP", *COLING '00 Proceedings of the 18th Conference on Computational Linguistics*, Stroudsburg, PA, pp. 62–68, 2000.

[BOE 01] BOERSMA P., HAYES B., "Empirical tests of the Gradual Learning Algorithm", *Linguistic Inquiry*, vol. 32, pp. 45–86, 2001.

[BON 03] BONASTRE J.-F., BIMBOT F., BOË L.-J. *et al.*, "Person authentication by voice: A need for caution", *Eurospeech*, Geneva, pp. 33–37, available at: http://www.afcp-parole.org/doc/AFCP_SpLC_HotTopicsEurospeech03_final.pdf, 2003.

[BOS 97] VAN DEN BOSCH A., Learning to pronounce written words: a study in inductive language learning, PhD Dissertation, University of Maastricht, 1997.

[BOU 83] BOULAKIA G., "Phonosyntaxe du français", *Revue Internationale du Traitement Automatique du Langage*, vol. 24, pp. 24–63, 1983.

[BOU 98] BOUFADEN N., Analyse syntaxique robuste des textes de dialogues oraux, Thesis, Laval University, Québec, 1998.

[BOU 02] BOUSQUET-VERNHETTES C., Compréhension robuste de la parole spontanée dans le dialogue oral homme-machine: Décodage conceptuel stochastique, PhD Thesis, Paul Sabatier University-Toulouse-III, 2002.

[BOU 09] BOUKOUS A., *Phonologie de l'amazighe*, Institut Royal de la Culture Amazighe, Rabat, 2009.

[BRE 82] BRESNAN J. (ed.), *The Mental Representation of Grammatical Relations*, MIT Press, Cambridge, MA, 1982.

[BRI 93] BRILL E., A corpus-based approach to language learning, PhD Dissertation, University of Pennsylvania, 1993.

[BRI 95] BRILL E., "Transformation-based error-driven learning and natural language processing: a case study in part of speech tagging", *Computational Linguistics*, vol. 21, no. 4, pp. 543–565, 1995.

[BRO 91] BROWN P.F., LAI J.C., MERCER R.L., "Aligning sentences in parallel corpora", *Proceedings of ACL 92*, pp. 169–176, 1991.

[BUR 15] BURKHARDT F., CAMPBELL N., "Emotional speech synthesis", in CALVO R., D'MELLO S.K., GRATCH J. *et al.* (eds), *The Oxford Handbook of Affective Computing*, Oxford University Press, 2015.

[CAB 98] CARABALLO S., CHARNIAK E., "New figures of merit for best-first probabilistic chart parsing", *Computational Linguistics*, vol. 24, no. 2, pp. 275–298, 1998.

[CAL 89] CALLIOPE, *La parole et son traitement automatique*, Masson, Paris, 1989.

[CAN 10] CANDITO M.-H., CRABBÉ B., DENIS P., "Statistical French dependency parsing: Treebank conversion and first results", *Proceedings of LREC'10 Conference*, La Valletta, Malta, 2010.

[CAR 83] CARBONNEL J.G., HAYES, P.J., "Recovery strategies for parsing extragrammatical language", *American Journal of Computational Linguistics*, vol. 9, nos. 3–4, pp. 123–146, 1983.

[CAR 74] CARTON F., *Introduction à la phonétique du français*, Bordas, Paris, 1974.

[CAR 06] CARNIE A., *Syntax: A Generative Introduction*, Blackwell, Oxford, 2006.

[CAV 98a] CAVAZZA M., "An integrated parser for TFG with explicit tree typing", *Proceedings of the International TAG Workshop*, Philadelphia, 28–31 July 1998.

[CAV 98b] CAVAZZA M., "Synchronous TFG for speech translation", *Proceedings of the International TAG Workshop*, Philadelphia, 28–31 July 1998.

[CHA 93] CHARNIAK E., HENDRICKSON C., JACOBSON N. *et al.*, "Equations for part-of-speech tagging", *National Conference on Artificial Intelligence*, pp. 784–789, 1993.

[CHA 98] CHAPELLIER J.-C., RAJMANN M., "A generalized CYK algorithm for parsing stochastic CFG", *Proceedings of 1st Workshop on Tabulation in Parsing and Deduction (TAPD'98)*, pp. 133–137, 1998.

[CHA 09] CHANG P.-C., TSENG H., JURAFSKY D. *et al.*, "Discriminative reordering with Chinese grammatical relations features", *Proceedings of the Third Workshop on Syntax and Structure in Statistical Translation*, pp. 51–59, 2009.

[CHE 93] CHEN S.F., "Aligning sentences in bilingual corpora using lexical information", *ACL93*, Columbus, Ohio, pp. 1–16, 22–26 June 1993.

[CHE 99] CHEN J., SRINIVAS B., SHANKER K.V., "New models for improving supertag disambiguation", *Proceedings of the 9th EACL*, Bergen, Norway, pp. 188–195, 1999.

[CHE 14] CHEN D., MANNING C., "A fast and accurate dependency parser using neural networks", *Conference on Empirical Methods on Natural Language Processing EMNLP*, Doha, Qatar, pp. 740–750, 25–29 October 2014.

[CHO 56] CHOMSKY N., "Three models for the description of language", *IRI Transactions on Information Theory*, vol. 2, no. 3, pp. 113–124, 1956.

[CHO 63] CHOMSKY N., SCHÜTZENBERGER M.P., "The algebraic theory of context free languages", in BRAFFORT P., HIRSCHBERG D. (eds), *Computer Programming and Formal Languages*, North Holland, Amsterdam, 1963.

[CHO 68] CHOMSKY N., HALLE M., *The Sound Pattern of English*, Harper and Row, New York, 1968.

[CHO 70] CHOMSKY N., "Remarks on nominalization", in JACOBS R., ROSENBAUM P. (eds), *Reading in English Transformational Grammar*, Ginn, Waltham, 1970.

[CHO 81] CHOMSKY N., *Lectures on Government and Binding*, Foris, Dordrecht, 1981.

[CHO 91] CHOI S.S., SON D.J., KIM J.C., "Unification in unification-based grammar", *6th Japanese-Korean Joint Conference on Formal Linguistics*, available at: http://dspace.wul.waseda.ac.jp/dspace/bitstream/2065/11788/1/JK6-26-34.pdf, 1991.

[CHO 95] CHOMSKY N., *The Minimalist Program*, MIT Press, Cambridge, 1995.

[CHU 88] CHURCH K.W., "A stochastic parts program and noun phrase parser for unrestricted text", *Proceedings of the Second Conference on Applied Natural Language Processing ANLC*, Stroudsburg, PA, pp. 136–143, 1988.

[CLE 76] CLEMENTS G.N., *Vowel Harmony in Nonlinear Generative Phonology: An Autosegmental Model*, Indiana University Linguistics Club, Bloomington, IL, 1976.

[CLO 81] CLOCKSIN W.F., MELLISH C.S., *Programming in Prolog*, Springer-Verlag, Berlin, 1981.

[COL 78] COLMERAUER A., "Metamorphosis grammar", in BLOC L. (ed.), *Natural Language Communication with Computers*, Springer-Verlag, Berlin, 1978.

[COL 99] COLLINS M., Head-driven statistical models for natural language parsing, PhD Dissertation, University of Pennsylvania, 1999.

[COL 11] COLLOBERT R., WESTON J., BOTTOU L. *et al.*, "Natural Language Processing (Almost) from Scratch", *Journal of Machine Learning Research*, vol. 12, pp. 2493–2537, 2011.

[COR 97] CORI, M., DE FORNEL M., MARANDIN J.-M., "Parsing repair", in MITKOV R., NICOLOV N. (eds), *Recent Advances in Natural Language Processing*, John Benjamins Publishing Company, 1997.

[COR 99] CORE M., SCHUBERT L., "Speech repairs: a parsing perspective", *ICPhS Satellite Meeting on Disfluency in Spontaneous Speech*, Berkeley, CA, July pp. 48–53, 1999.

[COV 94] COVINGTON M., *Natural Language Processing for Prolog Programmers*, Prentice Hall, Englewood, NJ, 1994.

[CRE 95] CREISSELS D., *Éléments de syntaxe générale*, Presses Universitaires de France, Paris, 1995.

[CRO 96] CROCKER M.W., "Mechanisms for sentence processing", Research paper, The University of Edinburgh, Centre for Cognitive Science, November, 1996.

[CRY 71] CRYSTAL D., *Linguistics*, Penguin, Harmondsworth, 1971.

[CRY 91] CRYSTAL D., *A Dictionary of Linguistics and Phonetics*, 3rd ed., Blackwell, London, 1991.

[DAE 96] DAELEMANS W., ZAVREL J., BERCK P. *et al.*, "MBT: a memory-based part of speech tagger-generator", *Fourth Workshop on Very Large Corpora*, pp. 14–27, 1996.

[DAE 10] DAELEMANS W., VAN DEN BOSCH A., "Memory-based learning", in CLARCK A., FOX C., LAPPIN S. (eds), *The Handbook of Computational Linguistics and Natural Language Processing*, Wiley-Blackwell, Malden, MA, 2010.

[DAN 00] DANLOS L., ROUSSARIE L., "La génération automatique de textes", in PIERREL J.-M. (ed.), *Ingénierie des langues*, Hermès, Paris, 2000.

[DEN 13] DEN DIKKEN M. (ed.), *The Cambridge Handbook of Generative Syntax*, Cambridge University Press, 2013.

[DER 88] DEROSE S.J., "Grammatical category disambiguation by statistical optimization", *Computational Linguistics*, vol. 14, no. 1, pp. 31–39, 1988.

[DES 90] DE SMEDT K., KEMPEN G., "Segment Grammar a formalisme for incremental generation", in PARIS C. *et al.* (ed.), *Natural Language Generation and Computational Linguistics*, Kluwer Academic Publisher, Dordrecht, 1990.

[DES 03] DESMET M., HAMON S., LAVIEU B., "Les grammaires HPSG", *Linx* vol. 48, 2003.

[DEW 98] DEWE J., KARLGREN J., BRETAN I., Telia Research, Assembling a Balanced Corpus from the Internet, available at: http://eprints.sics.se/63/1/Dropjaw_korpus.html, 1998.

[DIC 45] DICE L.R., "Measures of the amount of ecologic association between species", *Ecology* vol. 26, no. 3, pp. 297–302, 1945.

[DOS 55] DOSTERT L.E., *The Georgetown-I.B.M. experiment*, Locke and Booth Collection, 1955.

[DOU 72] DOUGLAS JOHNSON C., *Formal Aspects of Phonological Description*, Mouton, The Hague, 1972.

[DUB 94] DUBOIS J., GIACOMO M., GUESPIN L. *et al.*, *Dictionnaire de linguistique et des sciences du langage*, Larousse, Paris, 1994.

[DUT 96] DUTOIT T., PAGEL V., "Le projet MBROLA : vers un ensemble de synthétiseurs vocaux disponibles gratuitement pour utilisation non-commerciale", *Actes des Journées d'Études sur la parole*, Avignon, pp. 441–444, 1996.

[DUT 97] DUTOIT T., *An Introduction to Text-to-Speech Synthesis*, Kluwer Academic Publishers, Dordrecht, 1997.

[DUT 00] DUTOIT T., "Introduction au traitement automatique de la parole", available at: http://tcts.fpms.ac.be/cours/1005-07-08/speech/parole.pdf, 2000.

[EAR 70] EARLEY J., "An efficient context-free parsing algorithm", *Communications of the ACM*, vol. 13, no. 2, pp. 94–102, 1970.

[EIS 97] EISNER J., "Efficient generation in primitive Optimality Theory", *35th Annual Meeting of the Association for Computational Linguistics, ACL*, pp. 313–320, 1997.

[EIS 00] EISNER J., SATTA G., "A Faster Parsing Algorithm for Lexicalized Tree-Adjoining Grammars", *5th Workshop on Tree-Adjoining Grammars and Related Formalisms (TAG+5)*, pp. 14–19, Paris, 25–27 May, 2000.

[ELM 90] ELMAN J.L., "Finding structure in time", *Cognitive Science*, vol. 14, no. 2, pp. 179–211, 1990.

[EST 90] ESTIVAL D., "Generating French with a reversible unification grammar", *13th International Conference on Computational Linguistics*, Helsinki, Finland, pp. 106–111, 1990.

[FIL 88] FILLMORE C.J., KAY P., O'CONNOR Mary Catherine, "Regularity and idiomaticity in grammatical constructions: the case of let alone", *Language*, vol. 64, pp. 501–538, 1988.

[FIR 48] FIRTH J.R., "Sounds and prosodies", *Transactions of the Philological Society*, pp. 127–52, 1948.

[FIR 57] FIRTH, J.R., *Papers in Linguistics: 1934–1951*, Oxford University Press, 1957.

[FLO 02] FLORIAN R., Fast transformation-based learning toolkit, Technical Report Johns Hopkins University, available at: http://nlp.cs.jhu.edu/%7Erflorian/fntbl/tbl-toolkit/node1.html, 02-07 2002.

[FOS 98] FOSLER-LUSSIER E., Markov models and hidden Markov models – a brief tutorial, Technical Report TR-98-041 International Computer Science Institute, 1998.

[FRA 92] FRAKES W.B., "Stemming algorithms", in FRAKES W.B. (eds), *Information Retrieval: Data Structures and Algorithms*, Prentice-Hall, Upper Saddle River, NJ, 1992.

[FRA 99] FRANCIS H.S., GREGORY M.L., MICHAELIS L.A., "Are lexical subjects deviant?", In CLS-99.

[FRA 12] FRANCEZ N., WINTNER S., *Unification Grammars*, Cambridge University Press, 2012.

[FUN 94] FUNG P., MCKEOWN K., "Aligning noisy parallel corpora across language groups: word pair feature matching by dynamic time warping", *Proceedings of the Association for Machine Translation (AMTA) in the Americas*, pp. 81–88, 1994.

[GAD 89] GADET F., *Le français ordinaire*, Armand Colin, Paris, 1989.

[GAL 93] GALE W.A., CHURCH K.W., "A program for aligning sentences in bilingual corpora", *Computational Linguistics* vol. 19, pp. 75–102, 1993.

[GAR 68] GARDE P., *L'Accent*, PUF, Paris, 1968.

[GAR 87] GARSIDE R., "The CLAWS word-tagging system", in GARSIDE R., LEECH G., SAMPSON G. (eds), *The Computational Analysis of English: A Corpus-based Approach*, Longman, London, 1987.

[GAV 00a] GAVALDA M., "SOUP: a parser for real-world spontaneous speech", *Sixth International Workshop on Parsing Technologies (IWPT-2000)*, Trento, Italy, February 2000.

[GAV 00b] GAVALDÀ M., Growing semantic grammars, PhD Dissertation, Language Technologies Institute, School of Computer Science, Carnegie Mellon University, 2000.

[GAZ 85] GAZDAR G., KLEIN E., PULLUM G.K. *et al.*, *Generalized Phrase Structure Grammar*, Blackwell Publishing, Oxford, and Harvard University Press, Cambridge, MA, 1985.

[GAZ 89] GAZDAR G., MELLISH C., *Natural Language Processing in Prolog*, Addison Wesley, Wokingham, 1989.

[GEE 83] GEE J.P., GROSJEAN F., "Performance structures: a psycholinguistic and linguistic appraisal", *Cognitive Psychology*, vol. 15, pp. 411–458, 1983.

[GER 90] GERDEMANN D., HINRICHS E.W., "Funetor-driven natural language generation with categorial-unification", *International Conference on Computational Linguistics*, pp. 145–150, 1990.

[GER 01] GERS F.A., SCHMIDHUBER J., "LSTM recurrent networks learn simple context free and context sensitive languages", *IEEE Transactions on Neural Networks*, vol. 12, no. 6, pp. 1333–1340, 2001.

[GEU 02] GEUTNER P., STEFFENS F., MANSTETTEN D., "Design of the Vico spoken dialog system: evaluation of user expectations by Wizard of Oz simulations", *Proceedings of LREC02 Conference*, Las Palmas, Spain, 2002.

[GOL 76] GOLDSMITH J.A., Autosegmental phonology, PhD Dissertation, Massachusetts Institute of Technology, Cambridge, 1976.

[GOL 90] GOLDSMITH J.A., *Autosegmental and Metrical Phonology*, Blackwell, Malden, 1990.

[GOL 03] GOLDBERG A.E., "Constructions: a new theoretical approach to language", *TRENDS in Cognitive Sciences*, vol. 7, no. 5 pp. 219–223, May, 2003.

[GOL 07] GOLDWATER S., GRIFFITHS T.L., "A fully Bayesian approach to unsupervised part-of-speech tagging", *Proceedings of the 45th Annual Meeting of the Association of Computational Linguistics*, pp. 744–751, June 2007.

[GOO 98] GOODMAN J., Parsing inside-out, PhD Dissertation, Harvard University, 1998.

[GRA 10] GRAF T., Logics of phonological reasoning, Master Thesis, University of California, Los Angeles, 2010.

[GRE 65] GREIBACH S., "A new normal-form theorem for context-free phrase structure grammars", *Journal of the ACM*, vol. 12, no. 11, 1965.

[GRE 81] GREENE B.B., RUBIN G.M., *Automatic Grammatical Tagging of English*, Department of Linguistics, Brown University Providence, RI, 1981.

[GRE 99] GREFENSTETTE G., "Light parsing as finite-state filtering", in KORNAI A. (ed.), *Extended Finite State Models of Language*, Cambridge University Press, 1999.

[GRE 11] GREEN S., DE MARNEFFE M.-C., BAUER J. *et al.*, "Multiword expression identification with tree substitution grammars: a parsing tour de force with French", *2011 Conference on Empirical Methods on Natural Language Processing EMNLP*, University of Edinburgh, 27–29 July 2011.

[GRI 65] GRIFFITHS T.V., PETRICK S.R., "On the relative efficiencies of context-free grammar recognizers", *Communications of the ACM*, vol. 8, no. 5, pp. 289–300, 1965.

[GRI 06] GRIES S.T., STEFANOWITSCH A. (eds), *Corpora in Cognitive Linguistics: Corpus-Based Approaches to Syntax and Lexis*, Mouton de Gruyter, Berlin and New York, 2006.

[GRO 96] GROSS G., *Les expressions figées en français: noms composés et autres locutions*, Éditions Ophrys, Paris, 1996.

[GRO 12] GROSS M., LENTIN A., *Introduction to Formal Grammars*, Springer Science & Business Media, 2012.

[GRU 95] GRUNE D., JACOBS C., *Parsing Techniques: A Practical Guide*, Ellis Horwood Limited, Chichester, 1995.

[GUS 07] GUSSENHOVEN C., "Intonation", in DE LACY P. (ed.), *Cambridge Handbook of Phonology*, Cambridge University Press, 2007.

[HAB 97] HABERT B., NAZARENKO A., SALEM A., *Les linguistiques de corpus*, Armand Colin, Paris, 1997.

[HAF 74] HAFER M., WEISS S., "Word segmentation by letter successor varieties", *Information Storage and Retrieval*, vol. 10, pp. 371–385, 1974.

[HAJ 98] HAJIC J., "Building a syntactically annotated corpus: the Prague dependency treebank", in HAJICOVA E. (ed.), *Issues of Valency and Meaning, Studies in Honor of Jarmila Panevova*, Charles University Press, pp. 12–19, 1998.

[HAL 93] HALLE M., MARANTZ A., "Distributed morphology and the pieces of inflection", in KENNETH HALE, S., KEYSER J. (eds), *The View from Building 20*, MIT Press, Cambridge, MA, 1993.

[HAR 99] HARDCASTLE W.J., LAVER J. (eds), *The Handbook of Phonetic Sciences*, Blackwell, Malden, 1999.

[HAS 07] HASAN F.M., UZZAMAN N., KHAN M., "Comparison of unigram, bigram, HMM and Brill's POS tagging approaches for some South Asian languages", in ELLEITHY K. (ed.), *Advances and Innovations in Systems, Computing Sciences and Software Engineering*, Springer, 2007.

[HAT 99] HATON J.-P., "Neural network for automatic speech recognition: a review in speech processing", in CHOLLET G. *et al.* (ed.), *Recognition and Artificial Neural Networks*, Springer, 1999.

[HAY 60] HAYS D.G., "Grouping and dependency theories", P-1910, RAND Corporation, 1960.

[HAY 09] HAYES B., *Introductory Phonology*, Wiley-Blackwell, Malden, 2009.

[HDU 84] HUDSON R., *Word Grammar*, Blackwell, Oxford, 1984.

[HEE 95] HEEMAN P., ALLEN J., "The train 93 dialogs, TRAINS Technical note94-2", University of Rochester Computer Science Department, available at: ftp:// ftp.cs.rochester.edu/pub/.../ai/94.tn2.Trains_93_dialogues.ps.gz, March, 1995.

[HEE 97] HEEMAN P.A., Speech repairs, intonational boundaries and discourse markers: modeling speakers' utterances in spoken dialog, PhD Dissertation, University of Rochester, 1997.

[HEI 09] HEINZ J., KOBELE G., RIGGLE J., "Evaluating the complexity of optimality theory", *Linguistic Inquiry*, vol. 40, pp. 277–288, available at: http://roa.rutgers. edu/files/968-0508/968-RIGGLE-0-0.PDF, 2009.

[HEN 94] HENDERSON J., Description based parsing in a connectionist network. PhD Thesis, University of Pennsylvania, 1994.

[HEN 01] HENDRIKS P., DE HOOP H., "Optimality theoretic semantics", *Linguistics and Philosophy*, vol. 24, no. 1, pp. 1–32, February 2001.

[HEN 10] HENDERSON J., "Artificial neural networks", in CLARK A., FOX C., LAPPIN S. (eds), *Handbook of Computational Linguistics and Natural Language Processing*, Wiley-Blackwell, Malden, 2010.

[HES 05] HESS M., KLENNER M., Optimality theory and computational linguistics: an overview, Research paper, University of Zurich, 2005.

[HIN 06] HINTON G.E., OSINDERO S., TEH Y., "A fast learning algorithm for deep belief nets", *Neural Computation*, vol. 18, pp. 1527–1554, 2006.

[HIN 12] HINTON G.E., NITISH S., KRIZHEVSKY A. *et al.*, "Improving neural networks by preventing co-adaptation of feature detectors", *CoRR*, abs/ 1207.0580, 2012.

[HIR 84] HIRSHBERG J., PIERREHUMBERT J., The Intonational Structuring of Discourse, Association for Computational Linguistics Stroudsburg, PA, 1984.

[HIR 98] HIRST D., DI CRISTO A., (eds), *Intonation Systems: A Survey of Twenty Languages*, Cambridge University Press, 1998.

[HOC 55] HOCKETT C.F., *A Manual of Phonology*, Waverly Press and Indiana University Publications in Anthropology and Linguistics, Baltimore, 1995.

[HOC 58] HOCKETT C.F., *A Course in Modern Linguistics*, MacMillan, New York, 1958.

[HOE 04] HOEY M., "Textual colligation: a special kind of lexical priming", *Language and Computers* vol. 1, no. 49, pp. 171–194, 2004.

[HOF 99] HOFSTADTER D., *Gödel, Escher, Bach: An Eternal Golden Braid*, Basic Books, 1999.

[HOP 82] HOPFIELD J.J., "Neural networks and physical systems with emergent collective computational abilities", *The National Academy of Sciences*, vol. 79, pp. 2554–2558, 1982.

[HOP 01] HOPCROFT J.E., MOTWANI R., ULLMAN J.D., *Introduction to Automata Theory Languages, and Computation*, Addison Wesley, 2001.

[HUD 00] HUDSON R., *Dependency Grammar*, Esslli Summer School, University of Birmingham, UK, 2000.

[HUD 10] HUDSON R., *An Introduction to Word Grammar*, Cambridge University Press, 2010.

[HUN 01] HUNSTON S., "Colligation, lexis, pattern, and text", in HOEY M., MIKE S., GEOFF T. (eds), *Patterns of Text: In Honour of Michael Hoey*, John Benjamins, Amsterdam and Philadelphia, PA, 2001.

[HUT 04] HUTCHINS J., "The first public demonstration of machine translation: the Georgetown-IBM system", *7th January 1954, 6th Conference of the Association for Machine Translation in the Americas*, AMTA 2004, Washington DC, 28 September–2 October 2004.

[IDS 06] IDSARDI W., "A simple proof that optimality theory is computationally intractable", *Linguistic Inquiry*, vol. 37, pp. 271–275, 2006.

[INK 90] INKEKLAS S., ZEC D. (eds), *Phonology Syntax Connection*, Chicago University Press, IL, 1990.

[IRO 61] IRONS E.T., "A syntax directed compiler for ALGOL 60", *Communications of the ACM*, vol. 4, pp. 51–55, 1961.

[JAC 77] JACKENDOFF R., ‾X *Syntax: A Study of Phrase Structure*, MIT Press, Cambridge, MA, 1977.

[JAK 61] JAKOBSON R., FANT C., GUNNAR M. *et al.*, *Preliminaries to Speech Analysis: the Distinctive Features and Their Correlates*, MIT Press, Cambridge, MA, 1961.

[JAR 13] JARDINE A., "Logic and the generative power of autosegmental phonology", in KINGSTON J., MOORE-CANTWELL C., PATER J. *et al.* (eds), *Supplemental Proceedings of the 2013 Meeting on Phonology*, 2013.

[JEL 76] JELINEK F., "Continuous speech recognition by statisical methods", *IEEE Proceedings*, vol. 64, no. 4, pp. 532–556, 1976.

[JEL 92] JELINEK F., LAFFERTY J.D., MERCER R.L., "Basic methods of probabilistic context free grammars", *Speech Recognition and Understanding, NATO ASI Series*, vol. 75, pp. 345–360, 1992.

[JOD 93] JODOUIN J.-F., Réseaux de neurones et traitement du langage naturel : étude des réseaux de neurones récurrents et de leurs représentations, PhD Thesis, University of Paris 11, 1993.

[JOD 94] JODOUIN J.-F., *Les réseaux de neurones, principes et définitions*, Hermès, Paris, 1994.

[JOH 83] JOHNSON-LAIRD P.N., *Mental Models: Towards a Cognitive Science of Language, Inference, and Consciousness*, Harvard University Press, 1983.

[JOH 98] JOHNSON M., "Left corner transforms and finite state approximations", *36th Annual Meeting of the Association for Computational Linguistics and 17th International Conference on Computational Linguistics*, COLING-ACL '98, 10–14 August 1998.

[JOH 11] JOHNSON K., *Acoustic and Auditory Phonetics*, Wiley-Blackwell, Malden, 2011.

[JON 11] JONES K.S., "Natural language processing: a historical review", *Artificial Intelligence Review*, pp. 1–12, available at: http://www.cl.cam.ac.uk/archive/ksj21/histdw4.pdf, 10 December 2011.

[JOS 75] JOSHI A., LEVY L.S., TAKAHASHI M., "Tree adjunct grammar", *Journal of Computer and System Science*, vol. 21, no. 2, 1975.

[JOS 99] JOSHI A., SCHABES Y., "Tree-adjoining grammars", Michigan State University, available at: http://www.cis.upenn.edu/~joshi/, 15 March 1999.

[JOS 03] JOSHI A., SARKAR A., "Tree adjoining grammars and its application to statistical parsing", in BOD R., SCHA R., SIMA'AN K. (eds), *Data-oriented Parsing*, CSLI, 2003.

[JUR 00] JURAFSKY D., MARTIN J., *Speech and Language Processing*, Prentice Hall, 2000.

[JUT 94] HERAULT J., JUTTEN C., *Réseaux Neuronaux et Traitement du Signal*, Hermès, Paris, 1994.

[KAD 04] KADDOURI-ALHAMAD G., *Introduction à la phonétique de l'arabe*, Editions Ammar, Amman, 2004.

[KAH 12] KAHANE S., "Why to choose dependency rather than constituency for syntax: a formal point of view", in APRESJAN J., L'HOMME M.-C., IOMDIN L. *et al.* (eds), *Meanings, Texts, and Other Exciting Things: A Festschrift to Commemorate the 80th Anniversary of Professor Igor A. Mel'čuk*, Languages of Slavic Culture, Moscow, 2012.

[KAI 99] KAISER E., JOHNSTON M., HEEMAN P.A., "PROFER: predictive, robust finite-state parsing for spoken language", *Proceedings of ICASSP*, vol. II, March 1999.

[KAP 83] KAPLAN R., BRESNAN J., "Lexical-functional grammar: a formal system for grammatical representation", in JOAN B. (ed.), *The Mental Representation of Grammatical Relations*, MIT Press, Cambridge, MA, 1983.

[KAP 94] KAPLAN R., KAY M., "Regular models of phonological rule systems", *Computational Linguistics*, vol. 20, no. 3, pp. 331–378, 1994.

[KAP 97] KAPLAN R., "Finite state technology", in COLE R. (ed.), *Survey of the State of the Art in Human Language Technology*, Cambridge University Press, 1997.

[KAR 97] KARTTUNEN L., CHANOD J.P., GREFENSTETTE G. *et al.*, "Regular expressions for language engineering", *Natural Language Engineering*, vol. 2, no. 4, pp. 1–24, 1997.

[KAR 01] KARTTUNEN L., BEESLEY K.R., "A short history of two-level morphology", *ESSLLI-2001*, Helsinki, available at: http://www.helsinki.fi/esslli/, 2001.

[KAR 05] KARTUNEN L., BEESLEY K.R., *Twenty-five Years of Finite State Morphology*, CSLU Publications, available at: http://web.stanford.edu/~laurik/publications/25YearsOfTwoLM.pdf, 2005.

[KAS 65] KASAMI T., An efficient recognition and syntax analysis algorithm for context-free languages, Technical report AFCRL-65-758, Air Force Cambridge Research Laboratory, Bedford, MA, 1965.

[KAY 67] KAY M., "Experiments with a powerful parser", *Proceedings of the 2nd International Conference on Natural Language Processing*, 23–25 August, Grenoble, France, 1967.

[KAY 83] KAY M., Unification grammar, Technical report, Xerox Palo Alto Research Center, Palo Alto, CA, 1983.

[KAY 85] KAYE J., LOWENSTAMM J., VERGNAUD J.-R., "The internal structure of phonological elements: a theory of charm and government", *Phonology Yearbook*, vol. 2, pp. 305–328, 1985.

[KER 96] KERBRAT-ORECCIONI C., *La Conversation*, Seuil, Paris, 1996.

[KIE 00] KIEFER B., KRIEGER H.-U., NEDERHOF, M.-J., "Efficient and robust HPSG parsing of word hypotheses graphs", in WAHLSTER W. (ed.), *Verbmobil: Foundations of Speech-to Speech Translation System*, Springer, Berlin, 2000.

[KIM 03] KIM J.-D., OHTA T., TATEISI Y. *et al.*, "GENIA corpus: a semantically annotated corpus for bio-textmining", *11th International Conference on Intelligent Systems for Molecular Biology, Brisbane, Australia*, 29 June–3 July, 2003.

[KIN 07] KINGSTON J., "The phonetics–phonology interface", in PAUL D.L. (ed.), *The Cambridge Handbook of Phonology*, Cambridge University Press, 2007.

[KIN 10] Tomi K., Haizhou L., "An overview of text-independent speaker recognition: from features to supervectors", *Speech Communication*, vol. 52, no. 1, pp. 12–40, 2010.

[KIP 82] Kiparsky P., "Lexical phonology and morphology", in Yang S. (ed.), *Linguistics in the Morning Calm*, Hanshin, Seoul, 1982.

[KIR 04] Kiraz G.A., *Computational Nonlinear Morphology with Emphasis on Semitic Languages*, Cambridge University Press, 2004.

[KLA 80] Klatt D., "Software for a cascade/parallel formant synthesizer", *Journal of the Acoustical Society of America*, vol. 67, pp. 13–33, 1980.

[KLE 56] Kleene, S.C., "Representation of events in nerve nets and finite automata", in Shannon C., McCarthy J., (eds) *Automata Studies*, Princeton University Press, 1956.

[KLE 03] Klein D., Manning C., "A* parsing: fast exact viterbi parse selection", *Proceedings of the North American Chapter of the Association for Computational Linguistics (NAACL)*, 2003.

[KOE 05] Koehn P., "Europarl: a parallel corpus for statistical machine translation", 10th *Machine Translation Summit*, Phuket, Thailand, 12–16 September, pp. 79–86, 2005.

[KOH 82] Kohonen T, "Self-organized formation of topologically correct feature maps", *Biological Cybernetics* vol. 43, no. 1, pp. 59–69, 1982.

[KOS 83] Koskenniemi K., "Two-level morphology: a general computational model for word-form recognition and production", Publication No. 11, Department of General Linguistics, University of Helsinki, 1983.

[KRE 97] Krenn B., Samuelson C., "The linguist's guide to statistics", available at: nlp.stanford.edu/fsnlp/dontpanic.pdf, 1997.

[KRO 01] Kromann M., "Optimality parsing and local cost functions in discontinuous grammar", *Electronic Notes of Theoretical Computer Science*, vol. 53, pp. 163–179, 2001.

[KUR 00] Kurdi M.-Z., "The semantic tree unification grammar: a new formalism for spoken language parsing", *6th International Conference on Spoken Language Processing ICSLP'00*, Beijing, China, 16–20 October 2000.

[KUR 02] Kurdi M.-Z., "Combining pattern matching and shallow parsing techniques for detecting and correcting spoken language extragrammaticalities", *2nd Workshop on Robust Methods in Analysis of Natural language Data Romand 2002*, Rome, Italy, pp. 94–97, 17 July 2002.

[KUR 03] KURDI M.-Z., Contribution à l'analyse du langage oral spontané, Thesis, University Joseph Fourier, Grenoble, France, 2003.

[LAD 01] LADEFOGED P., *A Course in Phonetics*, 4th ed., Heinle & Heinle, Boston, 2001.

[LAM 05] LAMIROY B., KLEIN J.R., "Le problème central du figement est le semi-figement", *Linx*, vol. 53, pp. 135–154, available at: http://linx.revues.org/271, 2005.

[LAV 97] LAVIE A., GLR*: a robust grammar-focused parser for spontaneously spoken language, PhD Dissertation, Carnegie Mellon University, Pittsburgh, PA, 1997.

[LAZ 94] LAZARD G., *L'Actance*, Presses Universitaires de France, Paris, 1994.

[LEE 10] LEE K.Y., HAGHIGHI A., BARZILAY R., "Simple type-level unsupervised POS tagging", *Proceedings of the Conference on Empirical Methods in Natural Language Processing EMNLP '10*, Stroudsburg, PA, pp. 853–861, 2010.

[LEG 01] LEGENDRE G., GRIMSHAW J., VIKNER S. (eds), *Optimality-Theoretic Syntax*, MIT Press, Cambridge, MA, 2001.

[LEG 12] LEGALLOIS D., "La colligation : autre nom de la collocation grammaticale ou autre logique de la relation mutuelle entre syntaxe et sémantique?", *Corpus*, vol. 11, pp. 31–54, 2012.

[LEN 00] LENZO K.A., BLACK A.W., "Diphone collection and synthesis", *6th International Conference of Spoken Language Processing ICSLP*, Beijing, China, pp. 306–309, 2000.

[LÉO 05] LÉON P., *Phonétisme et prononciations du français*, Armand Colin, Paris, 2005.

[LEV 03] LEVY R., MANNING C.D., "Is it harder to parse Chinese, or the Chinese Treebank?", *ACL03*, pp. 439–446, 2003.

[LEV 05] LEVY R., Probabilistic models of word order and syntactic discontinuity, PhD Dissertation, Stanford University, 2005.

[LIB 09] LIEBER R., *Introducing Morphology*, Cambridge University Press, 2009.

[LIC 94] LICKLEY R.L., Detecting disfluency in spontaneous speech, Ph.D. Dissertation, University Edinburgh, 1994.

[LOP 99] LOPEZ P., Analyse d'énoncés oraux pour le dialogue homme-machine à l'aide de grammaires lexicalisées d'arbres, Doctoral thesis, University of Nancy, 1999.

[LUZ 87] LUZZATI D., "DIALORS: un système de dialogue oral simule pour une tache restreinte", *XVIème JEP*, Hammamet, pp. 183–186, 1987.

[LUZ 95] LUZZATTI D., *Le dialogue verbal homme-machine : étude de cas*, Masson, Paris, 1995.

[MAG 94] MAGERMAN D., Natural language processing as statistical pattern recognition, PhD Dissertation, Stanford University, 1994.

[MAH 95] MAHESH K., Syntax-semantics interaction in sentence understanding, PhD Dissertation, Georgia Institute of Technology, 1995.

[MAH 02] MAHEO-LE COADIC M., MIMRAN R., POISSON-QUINTON S., *Grammaire expliquée du français*, Clé International, Paris, 2002.

[MAN 88] MANN W., THOMPSON S., "Rhetorical structure theory. Toward a functional theory of text organization", *Text*, vol. 8, no. 3, pp. 243–281, 1988.

[MAN 99] MANNING C., SCHÜTZE H., *Foundations of Statistical Natural Language Processing*, MIT Press, Cambridge, 1999.

[MAN 00] MANA F., MASSIMINO P., PACCHIOTTI A., "Using machine learning techniques for grapheme to phoneme transcription", *7th European Conference on Speech Communication and Technology Eurospeech*, Aalborg, Denmark, pp. 1915–1918, 2001.

[MAR 80] MARCUS M., *A Theory of Syntactic Recognition for Natural Language*, MIT Press, Cambridge, MA, 1980.

[MAR 93] MARCUS M., SANTORINI B., MARCINKIEWICZ M.A., "Building a large annotated corpus of English: the Penn Treebank", *Computational Linguistics*, vol. 19, no. 2, pp. 313–330, available at: https://catalog.ldc.upenn.edu/docs/LDC95T7/cl93.html, 1993.

[MAR 94] MITCHELL M., KIM G., MARCINKIEWICZ M.A. *et al.*, "The Penn Treebank: annotating predicate argument structure", *Proceedings of the Human Language Technology Workshop*, San Francisco, available at: https://catalog.ldc.upenn.edu/docs/LDC95T7/arpa94.html, March, 1994.

[MAR 96] MARQUES N.C., PEREIRA LOPES G., "Using neural nets for Portuguese part-of-speech tagging", *5th International Conference on the Cognitive Science of Natural Language Processing*, pp. 21–22, 1996.

[MAR 06] MARNEFFE M.-C., MacCARTNEY B., MANNING C.D., "Generating typed dependency parses from phrase structure parses", *5th International Conference on Language Resources and Evaluation*, Genoa, Italy, 24–26 May 2006.

[MAR 07] MARCHAL A., *From Speech Physiology to Speech Phonetics*, ISTE, London and John Wiley & Sons, New York, 2007.

[MAR 08] MARTIN P., *Phonétique acoustique : Introduction à l'analyse acoustique de la parole*, Armond Colin, Paris, 2008.

[MAR 13] YUVAL M., HABASH N., RAMBOW O., "Dependency parsing of modern standard Arabic with lexical and inflectional features", *Computational Linguistics*, vol. 39-1, pp. 161–194, 2013.

[MAR 14] MARQUIS P., PAPINI O., PRADE H., *Panorama de l'Intelligence Artificielle: ses bases méthodologiques, ses développements*, Cépaduès Editions, Toulouse, 2014.

[MAY 95] MAYFIELD L., GAVALDA M., SEO, Y.-H. *et al.*, "Parsing real input in JANUS: a concept based approach", *Proceedings of TMI95*, pp. 442–447, 1995.

[MCC 43] MCCULLOCH W., PITTS, W., "Logical calculus of the ideas immanent in nervous activity", *Philosophy of Science*, vol. 10, no. 1, pp. 18–24, 1943.

[MCC 81] MCCARTHY J., "A prosodic theory of non-concatenative morphology", *Linguistic Inquiry*, vol. 12, pp. 373–418, 1981.

[MCE 96] MCENERY T., WILSON A., *Corpus Linguistics*, Edinburgh University Press, 1996.

[MCG 76] MCGURK H., MacDONALD J., "Hearing lips and seeing voices", *Nature*, vol. 264, no. 5588, pp. 746–748, 1976.

[MCK 98] MCKELVIE D., "The syntax of disfluency in spontaneous spoken language", HCRC Research Paper HCRC/RP-95, May 1998.

[MEG 03] MEGERDOOMIAN K., "Text mining, corpus building and testing", in FARGHALI A. (ed.), *Handbook for Language Engineers, Center for the Study of Language and Information*, Stanford University Press, 2003.

[MEL 88] MEL'CUK IGOR A., *Dependency Syntax: Theory and Practice*, SUNY Press, Albany, NY, 1988.

[MEL 99] MELAMED I.D., "Bitext maps and alignments via pattern recognition", *Computational Linguistics*, vol. 25, no. 1, pp. 107–130, 1999.

[MES 89] MESTER R.A., ITÔ J., "Feature predictability and underspecification: palatal prosody in Japanese mimetics", *Language*, vol. 65, pp. 258–293, 1989.

[MEY 04] MEYER C., *English Corpus Linguistics: An Introduction*, Cambridge University Press, 2004.

[MIL 04] MILTSAKAKI E., PRASAD R., JOSHI A. *et al.*, "The Penn discourse Treebank", *Proceedings of the NAACL/HLT Workshop on Frontiers in Corpus Annotation*, 2004.

[MIN 96] MINKER W., BENNACEF S., "Compréhension et évaluation dans le domaine ATIS", *Proceedings of the Journées d'Études en Parole*, 1996.

[MIS 87] MISRI G., Le figement linguistique en français contemporain, Linguistics Doctoral Thesis, University Paris V, 1987.

[MON 09] MONTAVON G., "Deep learning for spoken language identification", *NIPS Workshop on Deep Learning for Speech Recognition and Related Applications*, available at: http://research.microsoft.com/en-us/um/people/dongyu/nips2009/papers/montavon-paper.pdf, 2009.

[MOO 02] MOORE R.C., "Fast and accurate sentence alignment of bilingual corpora", in *Machine Translation: From Research to Real Users*, Springer-Verlag, Heidelberg, Germany, 2002.

[MOO 04] MOORE R.C., "Improved left-corner chart parsing for large context-free grammars", in BUNT H., CARROLL J., SATTA G. (eds), *New Developments in Parsing Technology*, Kluwer Academic Publishers, Norwell, MA, 2004.

[MOR 92] MOREL M.-A., Structure hiérarchique de l'énoncé oral, *International Symposium Lucien Tesnière*, Mont-Saint-Aignan, 19–21 November 1992.

[MUL 08] MULLER C., "Réflexions sur l'ordre des mots en français: les constituants majeurs de l'énoncé", in DURAND J., HABERT B., BERNARD L. (eds), *Proceedings of the 1st World Congress of French Linguistics, CD-Rom*, EDP Sciences, Paris, pp. 2663–2676, July 2008.

[NAK 94] NAKATANI C., HIRSHBERG J., "A corpus-based study of repair cues in spontaneous speech", *Journal of the Acoustical Society of America*, vol. 95, p. 160, 1994.

[NAS 09] NASR A., RAMBOW O., "Nonlexical chart parsing for TAG", in BANGALORE S., JOSHI A. (eds), *Complexity of Lexical Descriptions and its Relevance to Natural Language Processing: A Supertagging Approach*, MIT Press, Cambridge, 2009.

[NED 93] NEDERHOF M.-J., "Generalized left-corner parsing", *Proceedings of the Sixth Conference of the European Chapter of the Association for Computational Linguistics*, Utrecht, The Netherlands, pp. 305–314, 1993.

[NEL 64] NELSON F.W., "A standard sample of present-day english for use with digital computers", Internal Report, U.S Office of Education on Cooperative Research Project No. E–007, Brown University, Providence, 1964.

[NES 05] NESSELHAUF N., *Collocations in a Learner Corpus*, John Benjamins Publishing Company Philadelphia, PA, 2005.

[NES 07] NESPOR M., VOGEL I., *Prosodic Phonology*, Walter de Gruyter, Berlin, 2007.

[NIE 14] NIELSEN M., *Neural Networks and Deep Learning*, available at: http://neuralnetworksanddeeplearning.com/index.html, December 2014.

[OFL 94] OFLAZER K., GÖÇMEN E., BOZSAHIN C., *An Outline of Turkish Morphology*, available at: http://www.academia.edu/7331476/An_Outline_of_ Turkish_Morphology, October 1994.

[OGU 14] OGUNFUNMI T., NARASIMHA M., TOGNERI R. (eds), *Speech and Audio Processing for Coding, Enhancement and Recognition*, Springer, New York, 2014.

[OZK 94] OZKAN N., Analyses communicationnelles de dialogues finalisés, Thesis, Institute National Polytechnique de Grenoble, 1994.

[PAL 05] PALMER M., DANIEL G., KINGSBURY P., "The proposition bank: an annotated corpus of semantic roles", *Computational Linguistics*, vol. 31, no. 1, pp. 71–106, 2005.

[PAL 06] PALO P., A review of articulatory speech synthesis, Master's Thesis, Helsinki University of Technology, 2006.

[PAL 10] PALMER M., XUE N., "Linguistic annotation", in CLARK A., FOX C., LAPPIN S. (eds), *The Handbook of Computational Linguistics and Natural Language Processing*, Wiley-Balckwell, Malden, MA, 2010.

[PAR 93] PARADIS C., "Phonologie générative multilinéaire", in NESPOULOUS J.-L. (ed.), *Tendances actuelles en linguistique générale*, Delachaux and Niestlé, Paris, pp. 11–45, 1993.

[PAR 98] PARTINGTON A., *Patterns and Meanings: Using Corpora for English Language Research and Teaching*, John Benjamins, Amsterdam and Philadelphia, PA, 1998.

[PEC 04] PECMAN M., Phraséologie contrastive anglais-français: analyse et traitement en vue de l'aide à la rédaction scientifique, Doctoral Thesis, University of Nice-Sophia Antipolis, 2004.

[PEN 00] PENG L., "Nasal harmony in three South American languages", *International Journal of American Linguistics*, vol. 66, no. 1, pp. 76–97, 2000.

[PER 80] PEREIRA F., WARREN D., "Definite clause grammars for language analysis a survey of the formalism and a comparison with Augmented Transition Networks", *Artificial Intelligence*, vol. 13, pp. 231–277, 1980.

[PER 81] PEREIRA F., "Extraposition grammars", *Computational Linguistics*, vol. 7, no. 4, pp. 243–256, 1981.

[PIC 75] PICABIA L., *Éléments de linguistique générative: application au français*, Armond Colin, Paris, 1975.

[POE 04] POESIO M., "Discourse annotation and semantic annotation in the GNOME corpus", *Proceedings of the ACL Workshop on Discourse Annotation*, pp. 72–9, 2004.

[POL 87] POLLARD C., SAG I., *Information Based Syntax and Semantics*, Stanford, CSLI Publications, 1987.

[POL 96] POLLARD C., SAG I., "HPSG: background and basics", available at: http://www-users.york.ac.uk/~sjh1/courses/l231intro_to_hpsg/papers/hpsg-overview.pdf, 1996.

[POL 97] POLLARD C., SAG I., "HPSG: background and basics", in ABEILLE A. et al. (eds), *The Major Syntactic Structures of French, Esslli Summer School*, Aix-en-Provence, France, 1997.

[POL 98] POLLOCK J.-Y., *Langage et cognition: Introduction au programme minimaliste de la grammaire générative*, Presses Universitaires de France, Paris, 1998.

[POR 80] PORTER M.F., "An algorithm for suffix stripping", *Program*, vol. 14, no. 3, pp. 130–137, 1980.

[POR 06] PORTER M.F., Stemming Algorithms for Various European Languages, available at: http://www.snowball.tartarus.org/texts/stemmersoverview.html, 2006.

[PRI 93] PRINCE A., SMOLENSKY P., Optimality theory: constraint interaction in generative grammar, Technical Report, Rutgers University Center for Cognitive Science, 1993.

[PRI 04] PRINCE A., SMOLENSKY P., *Optimality Theory: Constraint Interaction in Generative Grammar*, Blackwell Publishing, Malden, 2004.

[PUS 03] PUSTEJOVSKY J., HANKS P., SAURI R. et al., "The timebank corpus", *Corpus Linguistics*, vol. 40, pp. 647–56, 2003.

[PUS 12] PUSTEJOVSKY J., STUBBS A., *Natural Language Annotation for Machine Learning*, O'Reily, Sebastopol, CA, 2012.

[RAB 89] RABINER L.R., "A tutorial on hidden Markov models and selected applications in speech recognition", *Proceedings of the IEEE*, pp. 257–286, 1989.

[RAP 11] RAPHAEL, L.J., BORDEN G.J., HARRIS K.S., *Speech Science Primer: Physiology, Acoustics, and Perception of Speech*, Wolters Kluwer/Lippincott Williams and Wilkins, Philadelphia, 2011.

[REI 04] REITTER D., "A development environment for multimodal functional unification generation grammars", *3rd International Conference on Natural Language Generation*, Brockenhurst, UK, 2004.

[REI 10] REITER E., "Natural language generation", in CLARK A., FOX C., LAPPIN S. (eds), *The Handbook of Computational Linguistics and Natural Language Processing*, Wiley-Blackwell, Malden, MA, 2010.

[REN 07] RENOUF A., BANERJEE J., "Lexical repulsion between sense-related pairs", *International Journal of Corpus Linguistics*, vol. 12, no. 3, pp. 415–444, 2007.

[RES 92a] RESNIK P., "Left-corner parsing and psychological plausibility", *International Conference on Computational Linguistics COLING'92*, Nantes France, pp. 191–197, 1992.

[RES 92b] RESNIK P., "Probabilistic tree-adjoining grammar as a framework for statistical natural language processing", *Proceedings of the Fourteenth International Conference on Computational Linguistics (COLING'92)*, Nantes, France, pp. 418–424, 1992.

[ROA 02] ROACH P., A little encyclopaedia of phonetics, available at: www.personal.rdg.ac.uk/~llsroach/encyc.pdf, 2002.

[ROA 07] ROARK B., SPROAT R., *Computational Approaches to Morphology and Syntax*, Oxford University Press, 2007.

[ROB 02] ROBERGE Y., Une brève introduction aux concepts de la syntaxe générative, online courses, University of Toronto, available at: http://french.chass.utoronto.ca/fre378/messages.html, 2002.

[ROC 96] ROCHE E., SCHABES Y., Introduction to finite-state devices in natural language processing, Technical Report, Mitsubishi Research Lab. TR1996-013, 1996.

[ROS 70] ROSENKRANTZ D.J., LEWIS P.M., "Deterministic left corner parsing", *IEEE Conference Record of the 11th Annual Symposium on Switching and Automata Theory*, pp. 139–152, 1970.

[ROS 81] ROSSI M., DI CRISTO A., HIRST D. *et al.*, *L'intonation : de l'acoustique à la sémantique*, Klincksieck, Paris, 1981.

[ROS 00] ROSSI M., *L'intonation, le système du français: description et modélisation*, Ophrys, Paris, 2000.

[ROS 05] ROSENBLUM L.D., "Primacy of multimodal speech perception", in PISONI D.B., REMEZ R.E. (eds), *The Handbook of Speech Perception*, Blackwell Publishing, Malden, MA, 2005.

[ROU 99a] ROUSSEL D., Intégration de prédictions linguistiques issues d'applications à partir d'une grammaire d'arbres hors contexte : contribution à l'analyse de la parole, Thesis, INPG, Grenoble, France, 1999.

[ROU 99b] ROUSSEL D., KURDI M.-Z., CAELEN J., "Normalisation des extragrammaticalités, supertagging et analyse partielle pour le traitement de la parole", *Hybrid Methods Workshop NLP/TALP for robust language processing*, Cargèse, 11–17 July 1999.

[ROU 00] ROUILLARD J., Hyperdialogue sur Internet: Le système HALPIN, Doctoral Thesis, University Grenoble I, 2000.

[RUM 86] RUMELHART D.E., McCLELLAND J.L., PDP RESEARCH GROUP, *Parallel Distributed Processing, Volume 1: Explorations in the Microstructure of Cognition: Foundations*, MIT Press, Cambridge, MA, 1986.

[RUS 10] RUSSEL S., NORVIG P., *Intelligence Artificielle*, 3rd ed., Pearson, 2010.

[SAB 83] SABAH G., RADY M., "A deterministic syntactic-semantic parser applied to French", *Actes 8 IJCAI, Karlsruhe*, pp. 707–710, 1983.

[SAG 03] SAG I.A., WASOW T., BENDER E.M., *Syntactic Theory: A Formal Introduction*, Center for the Study of Language and Information CSLI, Stanford, 2003.

[SAU 72] SAUVAGEOT A., *Analyse du français parlé*, Hachette, Paris, 1972.

[SCH 92] SCHABES Y., "Stochastic tree adjoining grammars", *Proceedings of the Fourteenth International Conference on Computational Linguistics (COLING'92)*, Nantes, France, pp. 140–145, 1992.

[SCH 94a] SCHMIDT H., "Part of speech tagging with neural networks", *Proceedings of the 15th International Conference on Computational Linguistics (COLING-94)*, 1994.

[SCH 94b] SCHMIDT H., "Probabilistic part-of-speech tagging using decision trees", *Proceedings of International Conference on New Methods in Language Processing*, Manchester, UK, 1994.

[SCH 95] SCHABES Y., WATERS R.C., "Tree insertion grammar: a cubic-time parsable formalism that lexicalizes context-free grammar without changing the trees produced", *Computational Linguistics*, vol. 21, no. 4, pp. 479–513, December 1995.

[SHA 93] SHABAN M., A minimal GB parser, BU-CS Technical report 39-013, Boston, 1993.

[SHI 86] SHIEBER S.M., *An Introduction to Unification-based Approaches to Grammar*, CSLI, 1986.

[SHI 87] SHIEBER S.M., "Separating linguistic analyses from linguistic theories", in WHITELOCK P. *et al.* (eds), *Linguistic Theory and Computer Applications*, Academic Press, Cambridge, 1987.

[SHI 90] SHIEBER S., SCHABES Y., "Synchronous tree-adjoining grammars", *13th International Conference on Computational Linguistics*, vol. 3, pp. 1–6, 1990.

[SHI 95] SHIMAZU H., TAKASHIMA Y., "Multimodal definite clause grammar", *Systems and Computers in Japan*, vol. 26, no. 3, pp. 93–102, 1995.

[SHR 94] SHRIBERG E., Preliminaries to a theory of speech disfluencies, PhD Dissertation, University of Berkeley, 1994.

[SIK 97] SIKKEL K., *Parsing Schemata – A Framework for Specification and Analysis of Parsing Algorithms*, Springer-Verlag, Berlin, 1997.

[SIL 15] SILBERZTEIN M., *Formalizing Natural Languages*, ISTE, London and John Wiley & Sons, New York, 2015.

[SIN 91] SINCLAIR J., *Corpus, Concordance, Collocation*, Oxford University Press, 1991.

[SLE 91] SLEATOR D., TEMPERLEY D., "Parsing English with a link grammar", Research report CMU-CS-91-196, 1991.

[SRI 99] SRINIVAS B., JOSHI A., "Supertagging: an approach to almost parsing", *Computational Linguistics*, vol. 20, no. 3, pp. 331–378, 1999.

[STE 88] STERIADE D., "Clements and Keyser: CV phonology", *Language*, vol. 64, pp. 118–130, 1988.

[TAY 09] TAYLOR P., *Text to Speech Synthesis*, Cambridge University Press, 2009.

[TES 59] TESNIÈRE L., *Éléments de syntaxe structurale*, Klincksieck, Paris, 1959.

[TES 98] TESAR B., SMOLENSKY P., "Learnability in optimality theory", *Linguistic Inquiry*, vol. 29, pp. 229–268, 1998.

[TES 12] TESAR B., "Learning phonological grammars for output-driven maps", *Proceedings of NELS 39*, p. 14, available at: http://roa.rutgers.edu/article/view/1043, 2012.

[THA 14] THAMPI S.M., GELBUKH A., MUKHOPADHYAY J. (eds), *Advances in Signal Processing and Intelligent Recognition Systems*, Springer International Publishing, Berlin, 2014.

[THO 68] THOMPSON K., "Programming techniques: regular expression search algorithm", *Communications of the ACM*, vol. 11, no. 6, pp. 419–422, 1968.

[TOM 86] TOMITA M., *Efficient Parsing for Natural Language*, Kluwer Academic Publishers, London, 1986.

[TOU 05] TOUTANOVA K., MANNING C.D., FLICKINGER D. *et al.*, "Stochastic HPSG parse disambiguation using the Redwoods corpus", *Research on Language and Computation*, vol. 3, no. 1, pp. 83–105, 2005.

[TRA 00] TRANEL B., "Aspects de la phonologie du français et la théorie de l'optimalité", *French Language*, vol. 126, pp. 39–72, 2000.

[TRO 69] TROUBETZKOY N., *Principles of Phonology*, University of California Press, Berkeley, CA, 1969.

[TRO 09] TROUILLEUX F., "Un analyseur de surface non déterministe pour le français", *Proceedings of the 16th Conference on Natural Language Processing (TALN'09)*, Senlis, 24–26 June 2009.

[TUR 50] TURING A., "Computing machinery and intelligence", *Mind*, vol. 59, no. 236, pp. 433–460, available at: http://loebner.net/Prizef/TuringArticle.html, October 1950.

[USZ 00] USZKOREIT H., KASPER W., FLICKINGER D. *et al.*, "Deep linguistic analysis with HPSG", in WAHLSTER W. (ed.), *Verbmobil: Foundations of Speech-to-Speech Translation*, Springer, Berlin, 2000.

[VAN 82] VAN DER HULST H.G., SMITH N., "An overview of autosegmental and metrical phonology", in VAN DER HULST H., SMITH N. (eds), *The Structure of Phonological Representations, Part I*, Foris, Dordrecht, 1982.

[VAU 00] VAUFREYDAZ D., BERGAMINI C., SERIGNAT J.-F. *et al.*, "A new methodology for speech corpora definition from internet documents", *Proceedings of the 2nd International Conference on Language Resources and Evaluation*, Athens, Greece, vol. 3, pp. 423–426, 2000.

[VIL 93] VILLARD P., "Morphologie: tendances actuelles", in NESPOULOUS J.-L. (ed.), *Tendances actuelles en linguistique générale*, Delachaux and Niestlé, Paris, 1993.

[VIT 67] VITERBI A.J., "Error bounds for convolutional codes and an asymptotically optimum decoding algorithm", *IEEE Transactions on Information Theory*, vol. 13, no. 2, pp. 260–269, 1967.

[VOG 09] VOGEL I., "Universals of prosodic structure", *Universals of Language Today: Studies in Natural Language and Linguistic Theory*, vol. 76, pp. 59–82, 2009.

[VOU 09] VOUTILAINEN A., "Part-of-speech tagging", in MITKOV R. (ed.), *The Oxford Handbook of Computational Linguistics*, Oxford University Press, 2009.

[WAR 91] WARD W., "Understanding spontaneous speech: the phoenix system", *Proceedings of International Conference on Acoustics, Speech and Signal Processing*, pp. 365–367, May 1991.

[WAT 02] WATSON J.C.E., *The Phonology and Morphology of Arabic*, Oxford University Press, 2002.

[WEB 04] WEBBER B., "D-LTAG: extending lexicalized TAG to discourse", *Cognitive Science*, vol. 28, no. 5, pp. 751–779, 2004.

[WEH 97] WEHRLI E., *L'analyse syntaxique des langues naturelles : problèmes et méthodes*, Masson, Paris, 1997.

[WIE 05] WIEBE J., THERESA-WILSON CARDIE C., "Annotating expressions of opinions and emotions in language", *Language Resources and Evaluations*, vol. 39, nos. 2–3, pp. 165–210, 2005.

[WIL 06a] WILLETT P., "The Porter stemming algorithm: then and now", *Program: Electronic Library and Information Systems*, vol. 40, no. 3, pp. 219–223, 2006.

[WIL 06b] WILSON A., ARCHER D., RAYSON P. (eds), *Corpus Linguistics around the World*, Rodopi, Amsterdam, 2006.

[WOO 70] WOODS Z.A., "Transition network grammar for natural language analysis", *CACM*, vol. 13, no. 10, pp. 591–606, 1970.

[XAV 05] XAVIER S.P.E., *Theory of Automata, Formal Languages and Computation*, New Age International Publishers, New Delhi, 2005.

[XIA 01] XIA F., PALMER M., "Converting dependency structures to phrase structures", *1st International Conference on Human-Language Technology Research*, pp. 61–65, 2001.

[YAN 03] YANNICK MATHIEU Y., "La Grammaire de Construction", *Revue des linguistes de l'université Paris Ouest Nanterre la défense*, vol. 48, pp. 43–56, 2003.

[YOU 67] YOUNGER D.H., "Recognition and parsing of context-free languages in time n^3", *Information and Control*, vol. 10, no. 2, pp. 189–208, 1967.

[YU 14] YU D., DENG L., *Automatic Speech Recognition: A Deep Learning Approach*, Springer, Berlin, 2014.

[ZUE 97] ZUE V., COLE R., WARD W., "Speech recognition", in COLE R. (ed.), *Survey of the State of the Art in Human Language Technology*, Cambridge University Press, 1997.

Index

S, T, V, X

Other titles from

in

Cognitive Science and Knowledge Management

2016

CLERC Maureen, BOUGRAIN Laurent, LOTTE Fabien
Brain–Computer Interfaces 1: Foundations and Methods
Brain–Computer Interfaces 2: Technology and Applications

FORT Karën
Collaborative Annotation for Reliable Natural Language Processing

GIANNI Robert
Responsibility and Freedom
(Responsible Research and Innovation Set – Volume 2)

LENOIR Virgil Cristian
Ethical Efficiency: Responsibility and Contingency
(Responsible Research and Innovation Set – Volume 1)

MATTA Nada, ATIFI Hassan, DUCELLIER Guillaume
Daily Knowledge Valuation in Organizations

NOUVEL Damien, EHRMANN Maud, ROSSET Sophie
Named Entities for Computational Linguistics

SILBERZTEIN Max
Formalizing Natural Languages: The NooJ Approach

2015

LAFOURCADE Mathieu, JOUBERT Alain, LE BRUN Nathalie
Games with a Purpose (GWAPs)

SAAD Inès, ROSENTHAL-SABROUX Camille, GARGOURI Faïez
Information Systems for Knowledge Management

2014

DELPECH Estelle Maryline
Comparable Corpora and Computer-assisted Translation

FARINAS DEL CERRO Luis, INOUE Katsumi
Logical Modeling of Biological Systems

MACHADO Carolina, DAVIM J. Paulo
Transfer and Management of Knowledge

TORRES-MORENO Juan-Manuel
Automatic Text Summarization

2013

TURENNE Nicolas
Knowledge Needs and Information Extraction: Towards an Artificial Consciousness

ZARATÉ Pascale
Tools for Collaborative Decision-Making

2011

DAVID Amos
Competitive Intelligence and Decision Problems

LÉVY Pierre
The Semantic Sphere: Computation, Cognition and Information Economy

LIGOZAT Gérard
Qualitative Spatial and Temporal Reasoning

PELACHAUD Catherine
Emotion-oriented Systems

QUONIAM Luc
Competitive Intelligence 2.0: Organization, Innovation and Territory

2010

ALBALATE Amparo, MINKER Wolfgang
Semi-Supervised and Unsupervised Machine Learning: Novel Strategies

BROSSAUD Claire, REBER Bernard
Digital Cognitive Technologies

2009

BOUYSSOU Denis, DUBOIS Didier, PIRLOT Marc, PRADE Henri
Decision-making Process

MARCHAL Alain
From Speech Physiology to Linguistic Phonetics

PRALET Cédric, SCHIEX Thomas, VERFAILLIE Gérard
Sequential Decision-Making Problems / Representation and Solution

SZÜCSAndras, TAIT Alan, VIDAL Martine, BERNATH Ulrich
Distance and E-learning in Transition

2008

MARIANI Joseph
Spoken Language Processing

Printed and bound by CPI Group (UK) Ltd, Croydon, CR0 4YY

27/10/2024

14580728-0003